The Quantum World
and the
Expansion of the Universe

Cosmological Model by Vortices

Sergio Antonio Meneghetti

Lior Shamir

Pindamonhangaba – São Paulo – Brazil
Kansas – United States
October 2024

Sergio Antonio Meneghetti – Lior Shamir

Cover: Sergio Antonio Meneghetti

Editing: Sergio Antonio Meneghetti

ISBN: 978-65-01-19793-7

Editorial label: Independently published
Image credits: Pixabay and Adobe Exprexx

Dados Internacionais de Catalogação na Publicação (CIP)
(Câmara Brasileira do Livro, SP, Brasil)

```
Meneghetti, Sergio Antonio
   The quantum world and the expansion of the
universe : cosmological model by vortices / Sergio
Antonio Meneghetti, Lior Shamir. -- Pindamonhangaba,
SP : Ed. dos Autores, 2024.

   ISBN 978-65-01-19793-7

   1. Astronomia 2. Buracos negros (Astronomia)
3. Estrelas 4. Galáxias 5. Hidrogênio 6. Planetas
I. Shamir, Lior. II. Título.
```

24-234286   CDD-520

Índices para catálogo sistemático:

1. Astronomia   520

Eliane de Freitas Leite - Bibliotecária - CRB 8/8415

To the reader:

The ceiling of knowledge.

This is one of the great mysteries of the universe, not only of humanity on Earth.

What makes life exciting is that knowledge is limitless, because, even with all the power of information, new facts are born every fraction of a second.

This book, written by Lior Shamir and Sergio Antonio Meneghetti, brings new information to the scientific world. Because it is new and bold, it will naturally open a new window on subjects that are so researched and debated.

The work will be divided into two parts. In the first part, Sergio Antonio Meneghetti will show the possible mechanisms of the birth of matter and its construction, from the quantum world to the expansion of the universe.

In the second part, Dr. Lior Shamir will present his research on the macrocosm, the behavior of galaxies, and the expansion of the universe, based on data provided by the most advanced telescopes of today, such as James Webb and Hubble, on redshift.

Have a nice reading!

Summary

Introduction

Quantum mechanics has brought important information on cosmic construction.

Microscopic physics challenges the sharpest minds and reveals seemingly unpredictable mechanisms.

Due to the subtlety of particles and energy waves, the difficulty of more robust tests gives the impression of a world distinct and detached from the human scale and from the macrocosm.

What this book will bring to the reader and researcher is a new vision of these mechanics. It will unravel a guiding thread that will connect the microcosm with the macrocosm using the same principles of development of physical and chemical phenomena. An order that repeats itself from scale to scale.

When these new concepts crystallize in the minds of researchers, science will certainly make unimaginable advances. This knowledge will give humanity new technologies and, most importantly, new morals and ethics.

In relation to the macrocosm, a model of expansion will be presented that differs from the model imagined or deduced by current science. The volumetric linear expansion model will give way to the vortex expansion model.

This information will be presented in a logical manner and with references already researched by science.

This is not the author's guesswork or just scientific speculation: the basis lies in the elementary properties of each phenomenon and is intrinsic to the behavior of structures, from particles to complete atomic structuring.

The picture presented is peculiar in terms of movements and the fruits of these movements. Universal dynamics are governed by Laws and, therefore, unalterable throughout periods of universal development.

Such facts should be explored by minds seeking new possibilities in science as a whole.

The challenges are great and the result will be worth all the efforts and boldness.

With this information, many questions can be answered in a logical manner demonstrating the progression of the principles of the mechanisms that govern the universe.

Great powers related to nature require great responsibility.

Man will not only learn more about natural mechanisms, but will also have his intellectual and psychic self-development, giving him greater powers.

There are ways of learning beyond the classic models of how to obtain knowledge, and what will be presented in the first part of this book will be supported in part by these possibilities. One of them is through Synthetic Intuition (psychic insights).

Among the sources of information, Sergio Antonio Meneghetti based his knowledge on the work:

• The Great Synthesis by Pietro Ubaldi (1931 – 1935).

First Part

Sergio Antonio Meneghetti

8

Chapter 

Movements and Particles

Undoubtedly, without movement, there is no physical universe.

This is the crucial point for understanding the mechanisms of the universe, from the waves that form elementary particles to the expansion of the universe.

One of the great challenges of science is understanding these movements and especially how they promote the generation of solidity through energy, that is, how to concentrate energy for the formation of matter.

From this point on, I ask the researcher and reader to follow the logic of the physical phenomena presented in this work to evolve on the subject.

- How can waves be concentrated so that they generate solid particles?

To be able to provide answers, it will be necessary to understand a little about waves, since they are the basis of matter.

- What is the reason for this type of manifestation of energy in the form of waves?

When observing natural phenomena, we notice that they always use the exact amount of energy for their manifestations, that is, the process of involvement and development.

Something crucial in this context is that the nature of phenomena, whether physical or chemical, will always use the path of least energy consumption to perform any manifestation with its greatest yield.

With this information, it is deducible that, in the case of waves, this type of energy movement is the most efficient possible.

For the reader to be able to easily identify this concept, it is enough to remember climbing a steep path with a bicycle. If you go up the slope linearly, the physical effort will be great, but if you go up it by making curves to the right and left, despite the path being longer, the effort will be less.

Another example that can demonstrate this phenomenon is when using a toy called a roller skate, which consists of a front central bearing on an axle and two rear bearings at the ends. To be able to break the inertia on flat terrain, it was enough to move the axle to the right and to the left several times, for the toy to move out of place (inertia) and increase its speed, the more these movements were made.

Thus, as in the example of the roller-bearing toy car, the more energy I give to the movements of going right and left, the more linear speed, that is, moving forward, I will impose on the toy. Drawing a parallel, the more energy I put into the atom or molecule, the more excitement there will be in these bodies, thus, there will be greater vibration in the atoms or in the waves that make up the particles and the atomic nucleus.

It is known in science that the more energy or vibration a wave has, the sharper its bumps will be, the shorter the wave in length and the higher its speed. On the other hand, the longer the wave, the lower its vibration and speed.

In its particularities, each type of wave has its own quantum of energy, model, length, vibrational level and behavior.

With these observations, it is easier to do a mental exercise to be able to imagine the first question of how to concentrate energies.

Waves   Particles (example: Helium)

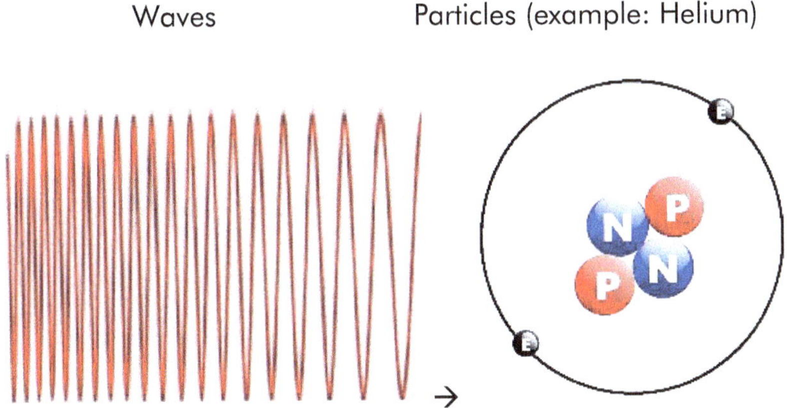

Figure 01- Wave heading towards its destination as matter

Above we have a simple wave model and a Helium atom exemplifying three types of particles: the electron, as a simple particle, the neutron and the proton, as particles composed of smaller particles.

- How can this type of energy behavior (wave) become a spherical model with characteristics of solidity?

- How can these waves acquire stability to withstand the condition of solidity in constant shocks in nature?

It is worth noting that waves are never inert.

Pause your reading and try to imagine how you would concentrate waves using the natural processes of the universe. To do this, try to disconnect from preconceived theories or hypotheses. Imagine that you are exploring the phenomena.

It may seem difficult, because the mind is saturated with information, hypotheses and theories are constantly being thrown around, and few dare to challenge them.

It is almost certain that most people would imagine that, at some point and for some unknown reason, an explosion or energetic expansion would cause waves to assume spherical characteristics.

Simplifying this possibility, we will imagine an explosion of energies.

The following drawing should demonstrate logically the behavior of the waves being thrown in all directions.

Evaluate the possibilities of these waves coming together through this process.

Figure 02 – Dispersion waves

- Would it be possible for these waves launched toward infinity to meet and concentrate?

If so, the question remains:

- What could be the physical condition of this encounter?

Starting from another hypothesis, such as the result of vibratory waves coming from strings.

- How would these vibrations, or how would these manifestations occur to concentrate at some point?

If this is the basis or origin of the movement of waves, it is known that waves always have vibratory behavior, as this is what gives them the conditions to move in a straight line in any direction. So, it is still difficult to imagine the point where these waves could meet to form something compact, stable and spherical.

It is important to keep in mind that the Law of Physics, in a phenomenon, will always have to remain unchanged, both now and in the distant past, that is, billions of years ago.

This information is to support the idea that, in the past, the Laws of Physics could not have behaved differently than they do today.

This disproves the idea that a physical condition, governed by the Law of Physics, manifests itself differently at different times. In this way, a Law would cease to be a Law.

How we can deduce, packaging or concentrating energies naturally is a great challenge.

Despite this apparent difficulty, it is enough to observe nature itself carefully, because it shows us all the time – and in the most varied events – the basic principles of the phenomena.

To observe microscopic or macroscopic conditions, these conditions require advanced and expensive technologies. The interesting thing would be to observe the conditions closest to our dimensional reality, that is, that which is in our size.

- Although Quantum Physics and Cosmology are apparently different in their manifestations, I dare say – and I will try to demonstrate through simple processes –that reality is different and that both have a common thread through basic principles.

- I want to make it clear that, between micro and macro, the limits are not there in the minimum or maximum, but in our conditions of observation and intellectual understanding.

- With this statement, I want to say that the horizons of the scientific past were much smaller than the horizons of today, and that, without a doubt, the horizon will expand as we advance.

- In other words, we cannot delude ourselves into wanting to have answers for everything. This applies especially to wanting to determine the probable beginning of the universe as we know it, as well as its time and existential destiny.

- The belief, even with scientific evidence, in a relative universe, that is, with a beginning, middle and end, can block us from advancing in understanding something possibly much greater than our intellectual reality.

Instead of asking the question:

- Where do we come from and where are we going?

The wisest question would be:

- Where and how are we coming from and where and how will we be going?

The connotation may seem the same, but the idea is to remove the imaginary limits of origin and destination.

This philosophical pause is necessary to open up our range of possibilities in this study.

Observing phenomena on our planet, or even outside it, we will pay attention to the natural phenomena of <u>concentration and expansion</u>, as this will be the focus of this section.

From the drain in your sink to tornadoes, waterspouts, quasars and black holes, this principle of Centripetal Force is the key to concentrating energy and matter naturally, that is, without human action.

If these natural phenomena occur before our eyes using the same principle, this is due to conditions that we know through research or, in some cases, the generating cause is still beyond our understanding.

With these comparisons, it is easier and more logical to launch hypotheses in the quantum world, as it gives us more security and support through facts.

If these phenomena occur in our easily observable world, that is, in a scale close to our size as human beings, we can deduce that the same phenomena can also occur in the microcosmic world.

Moving forward with this new possibility in the book, it is possible to propose the hypothesis that one of the models of energy concentration in the form of waves would be through dynamic concentration in the form of a vortex by Centripetal Force.

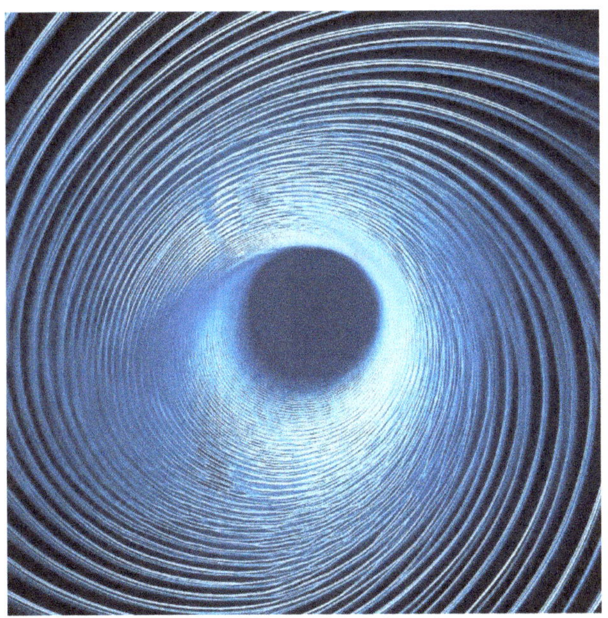

Figure 03 - Wave Concentration Vortex.

The other mode of concentration would be by wave dragging through the generation of the model mentioned above.

In the case of dynamic concentration by centripetal force, the waves would be constrained to a rotational funnel at very high speeds. In the second case, since there are no voids in the universe, this vortex movement generates attraction on its outer periphery, attracting whatever is within its reach and strength.

- Isn't this what happens in tornadoes?

In the following illustration, we can exemplify this possibility.

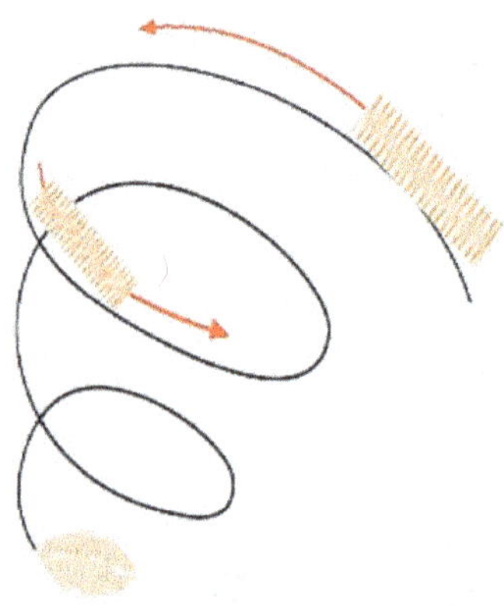

Figure 04 – Simplified model of Wave Concentration Vortex.

- - What causes this phenomenon in the quantum field?
- We still don't know.
- What we can say is that processes occur in the universe that are still mysteries to our intellectual and observational moment. This does not prevent deductions.
- - What would the mechanism of this concentration be like?
- The waves undergo a process similar to rolling a ball of wool, that is, a process where the waves would act like pieces of wool threads rolling over themselves at very high speeds, concentrating and generating the shape of dense spheres of energy, and, due to the forces and speeds exerted in this process, the set of waves would achieve the characteristic of Solidity.
- The resulting circular movement gives stability to the set.
- To be clearer, the formation of observable or detectable elementary particles.

With this process, **Matter** is born.

We are talking about movements and concentration of energies, but where do we get the raw material for such phenomena?

Common sense and logic are of great value in helping us reflect and dare to explore new concepts, even if they may be difficult to assimilate mentally.

Since there is no magic in the universe, everything happens through some primordial cause; certainly, nothing would ever produce anything. By the same reasoning, some existing thing will never be annihilated into nothing.

This is a fact.

With this universal fate, we can deduce that everything that exists now, in some way, existed in the past and will always exist in the future. The secret lies in the transformation of things.

With this concept, we can better work the mind to understand that we live in a <u>universal sea of substances in transformation</u>. And, regardless of whether or not we are aware of these manifestations, movement is the key to transformations. In other words, there is no inertia in the universe.

Another fact to be taken into account is that there can be no voids in the universe, that is, in some way, everything is occupied.

In our specific case, as human beings with five senses, we are deficient in perceiving what really exists, not to mention our evolutionary and intellectual moment. Even with all the technology available.

The reason for these clarifications is to broaden our vision of the universe. With a little imagination and logic, we can deduce that the dimensional limits (size of things) can move towards the less infinite ($-\infty$) and towards the more infinite ($+\infty$).

In our case, studying the quantum world, it is possible to imagine that there are substances, structures or processes, below the measurements already known to science.

These concepts are important for us to advance future hypotheses.

As a first case, it is to indicate the source from which dynamic energies arise or the basis for the formation of the first elementary particles that science was able to detect or deduce.

In a second case, to have a better idea of what the constitution of the universe is and how it works.

Always remember that with each new, more powerful instrument, the universe expands, so many concepts end up being in doubt.

Something that also limits us is imagining that we know all the phenomena and wanting to get all the answers from them.

Spatial dimensions, for example, have more measurable properties, while time, in addition to being more conceptual, is relative to each point in space and each observer.

Like everything we see or perceive through our senses or detections, these processes occur through the manifestations of waves throughout the universe. So, by deduction, time would be the period of transmission of the wave.

There is a controversial hypothesis that if I traveled at the speed of light, time would be equal to zero. Almost all of us are aware of this, but what is different to be extracted from this context is that if this were possible for human beings, their perceptions or sensations would be completely new, that is, a reality totally different from the one we know.

This example serves to give us an idea of existing conditions and realities, which we have no idea of what they are like. This is due to our personal and technological limitations.

Therefore, there is the possibility of dimensions prior to spatial dimensions and with manifestations that are totally different from our reality, such as other dimensions beyond those known with unimaginable manifestations.

With this apparent diversion from the subject, the message is that there must have been manifestations before the known quantum world. Thus, this can show that the universe is far beyond an initial point imagined or deduced by calculations and that it can follow different paths than those imagined by the scientific moment.

I believe that there is an Order, a perfect mechanism, that keeps everything in development and balance.

Why do I say this? Because I believe in the Laws of Physics known as immutable: all mechanisms of formation of matter until the expansion of the universe follow these Laws.

Returning to the main theme of this chapter, the particles produced by concentration vortices will be the first bricks of the great atomic construction.

If these particles were generated in this way, they are naturally rotating on their axis and this circular movement generates a field around them.

The formation of the nucleus, the basis of the atom, is the result of the synchronization of forces that, after growth, provide it with stability and repeatability. The latter gives neutrons identical structures to each other, as well as to protons and electrons with their respective counterparts.

- But what makes these particles bond or remain structured?

Of course, a perfect balance of forces.

The atomic nucleus is like a plant seed, in which resides all the potential of the future plant.

- There's a mechanism that can generate attraction and repulsion over these particles. What is it like?

Mechanics of Magnetism

Using the example of the terrestrial tornado and the circular motion of the particle, a parallel can be drawn on the smallest scales, as well as on the largest scales of the universe.

The vortex movement and the circular motion are the only models to promote the concentration, stability and expansion of small bodies, quantum systems and even gigantic systems, such as galaxies, black holes, and quasars, among others.

With a description, we will also be able to understand how the phenomena of attraction and repulsion occur, that is, the (-) and the (+).

The concentration vortex generates the attraction by the vortex model, attracting energies, or on a larger scale, mass. <u>Centripetal Force</u>.

Figure 05 – Land tornado.

Expansion vortex generates the repulsion of energies or, on a larger scale, mass. This is the Centrifugal Force.

With these two expressions, one can understand why negative attracts positive and why like charges repel each other. For there to be attraction, there must be complementary movements.

In Figures 4 and 5, we have two situations of attraction due to the action of concentration by centripetal force.

In the case of a galaxy, for example, we will have the movement of expansion, the action of centrifugal force.

To illustrate complementary movements, we will have three situations: two vortices generating repulsion because they are expansion vortices; two vortices generating repulsion because they are concentration vortices; and, finally, two vortices attracting each other because they are one expansion vortex and one concentration vortex, complementing each other. These phenomena can occur at the nuclear, orbital, or any other level.

A simple example to illustrate the phenomenon: two fans, one against the other, in ventilation mode (expansion), the air vortices repel each other; two fans, one against the other, but with one in ventilation mode (expansion) and the other in exhaust mode (concentration), there is the attraction of complementary air vortices.

Imagine that the air is a field or an electronic cloud.

All mass or manifestation in the universe, because they are products of movement, generate manifestations in their peripheries, from slight fluctuations to strong attractions or repulsion.

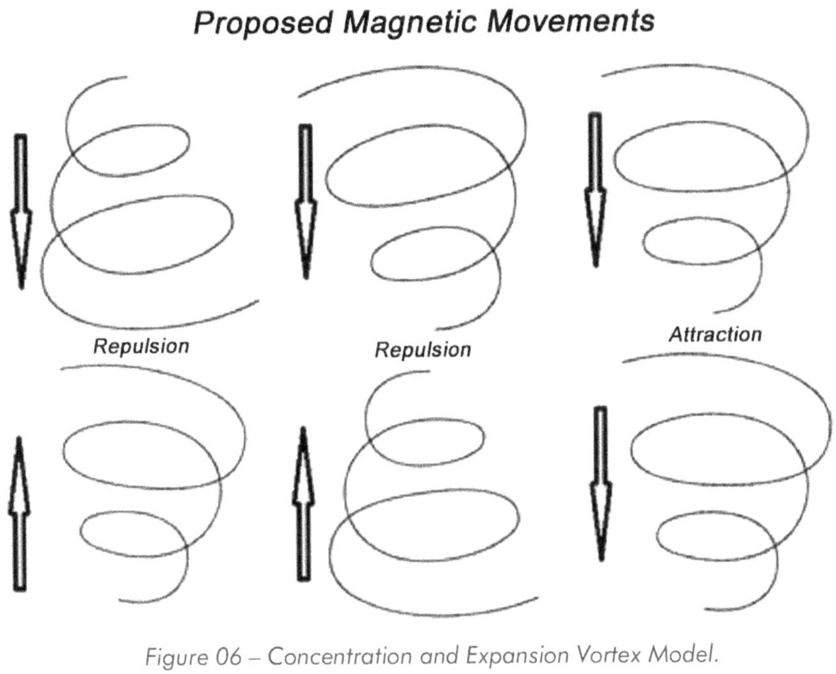

Proposed Magnetic Movements

Figure 06 – Concentration and Expansion Vortex Model.

It is important to note that the vortex-shaped movement model is what determines much of the behavior of physical and even chemical phenomena that exist in nature.

An important piece of evidence published in Nature Magazine on April 10, 2024, shows a comparison with the above hypothesis.

With the title **"Terahertz electric-field-driven dynamical multiferroicity in SrTiO₃"**, scientists generated small magnetism in neutral material through light in the form of a corkscrew, that is, in a kind of vortex-shaped movement.

More details and credit at the link:

https://www.nature.com/articles/s41586-024-07175-9

If we adapt these concepts to a particle, we can deduce that a minimum gravity can occur in this particle, due to the vortex movement that this particle has, thus generating a small field around it.

If we use this same concept for the nuclear system, that is, for a Neutron or Proton, the balance of forces will be obtained by these movements ordered like parts of a clock.

In the case of the Neutron, neutrality is achieved by the total balance between the charges of the particles that make up its "planetary system".

In the case of the Proton, there is a lack of a part with a negative characteristic to complement and give it neutrality. In this way, the Proton has a positive charge (+).

In order to remain solid spheres, it is important to emphasize that both neutrons and protons are, internally, similar to the atomic system. In other words, they are endowed with nuclei and particles that rotate around them, giving them solidity in the same way that the electron provides it to the atom.

Nuclear Physics has already detected several particles that make up these composite particles, namely the Neutron and the Proton.

Nuclear Physics has already detected several particles that make up these composite particles, namely the Neutron and the Proton.

Some observations that can be very useful in the case of discovery or detection processes in the world of particles:

- If a particle is a tangle of waves, that is, a roll of waves upon waves acquiring a spherical and circular shape, when these particles are thrown at very high speeds in a collider, it would be similar to two onions colliding. This comparison is the closest I have found to express the idea.
- When two onions collide with a lot of energy, they will break into several layers and in different sizes. In the case of particles, waves of energy will probably come out and pieces of particles that have still remained stabilized may remain.
- In the specific case of Neutrons and Protons, the rupture will scatter the constituent parts of the planetary system, that is, many of the parts already known to science and energy waves.
- The greater the energy and speed used in the collider, the greater the damage to the particles.
- Let's do another mental exercise:
- If we compare an elementary particle, a nuclear composite particle (Proton), an atom, a crystal, a star, a solar system and a galaxy, we can observe that, in all cases, there is repeatability of the growth process, whether by concentration or expansion.

Everything has an order that obeys the Laws of Physics or Chemistry.

Note that there is a growth in the systems of energy to form the elementary particle, of the sum and balance of particles to form the planetary system of a neutron, of the nucleus with the electron to form the atom, of the energies that promote atomic growth, of atoms that form crystals, of atoms that form stars, of stars that form solar systems, of solar systems that form galaxies and of galaxies that form systems of galaxies, etc.

If we compare these processes with the processes of life, both seem to follow the same guiding line: a process of growth that is organized and naturally determined.

With life, there is a need for nutrients to grow.

In the case of apparently inert matter, there is also a need for food. In the case of the atom, energies.

So far, I do not have a deeper understanding of the mechanism of the particles that make up the Neutron and the Proton. However, there is a logic to this process, and as reported, a construction where everything tends towards a balance of forces.

In each piece, for reasons that we do not yet fully understand, there is a process of growth, or, more profoundly, a becoming that coordinates all these ascending movements of the formation of matter as we know it.

Everything indicates that there is a guiding principle, an order that occurs in every corner of the universe, respecting the Laws of physics and chemistry. An invisible and abstract Essence that perfectly projects all constructions from the micro to the macrocosm.

It is important to emphasize that the theories regarding what is are not being modified, only hypotheses are being added about how the processes unfold.

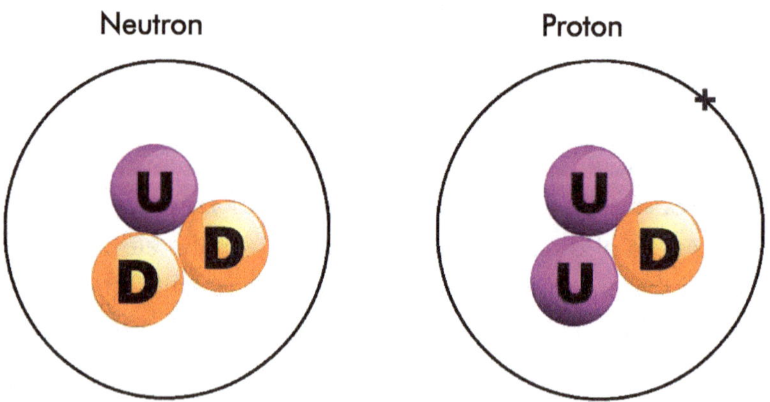

Figure 07 and 08 – Neutron and Proton Model.

Chapter II

Birth of Hydrogen

We know that Hydrogen, as the first and simplest element in the periodic table, is composed of a Proton as the nucleus and an Electron as the orbital particle.

It is widely known that the Proton has a positive charge (+) and that the Electron has a negative charge (-). Due to the balance of these two charges (+) and (-), the planetary system remains cohesive. One compensates for the other and this generates the stability of this element.

In the case of Hydrogen, there are two isotopes, Deuterium and Tritium.

Hydrogen

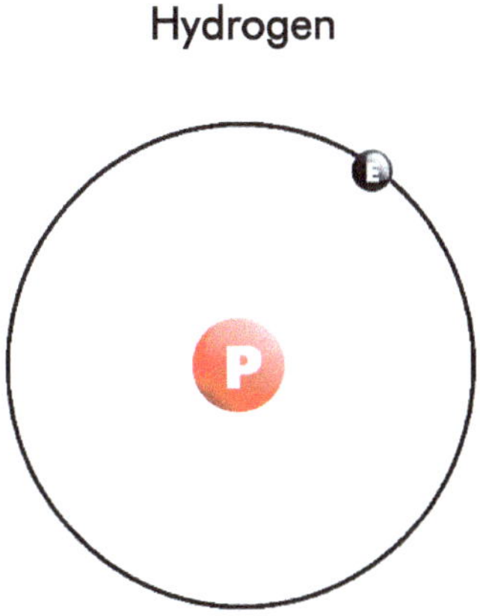

Deuterium is composed of a Proton, a Neutron and an Electron.

Deuterium

Tritium is composed of one Proton and two Neutrons.

Tritium

Although Deuterium has one Neutron, the equilibrium of the Hydrogen atom or isotope is maintained. Only the mass of the system increases and the smallest quantity in nature.

In the case of Tritium, with two Neutrons, the equilibrium is maintained, but tending towards imbalance due to one more Neutron. Similar to Deuterium, there is an increase in mass and it is rarer in nature. This process follows the nature of growth: an imbalance that is compensated for by a new equilibrium.

The structure of a Neutron, neutral, seems to have nothing in common with that of Hydrogen, since the latter is composed of two particles with different charges, but this is precisely the crucial point. Here, a very important hypothesis is inserted, that the growth or expansion of the Neutron will emit a particle that will be vital for the <u>birth of the atom</u>.

It can be imagined that, due to its very high speed inside the Neutron, some small particle could escape, remaining, however, attached to the system by the attraction of the charges (+) and (-).

Emission of an electron from within the Neutron.

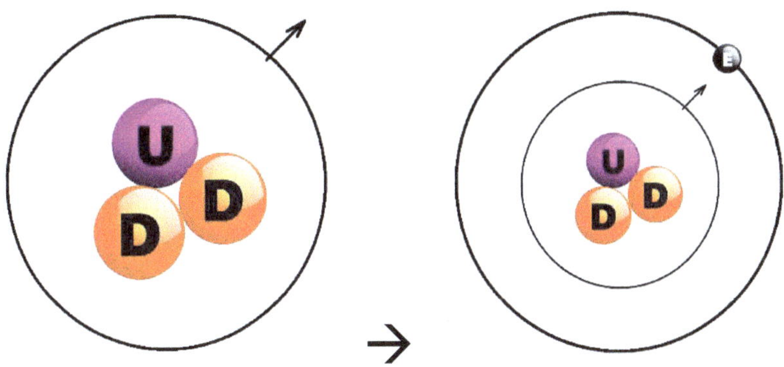

Transmutation of Neutron to Proton.

Neutron after emission = Formation of Proton

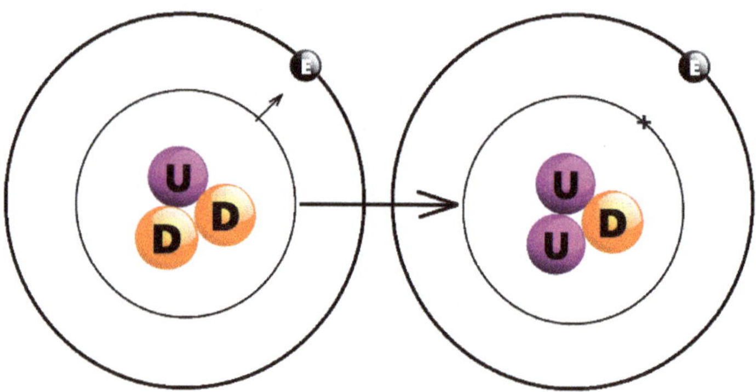

In this transition, it is observed that a Down Quark transforms into an Up Quark.

This change is already observed in atomic decay.

This process can occur with the imbalance of forces inside the Neutron, which, upon losing a negatively charged particle, changes its condition from neutrality to positivity, that is, from Neutron to Proton. In this way, Hydrogen is generated.

Hydrogen

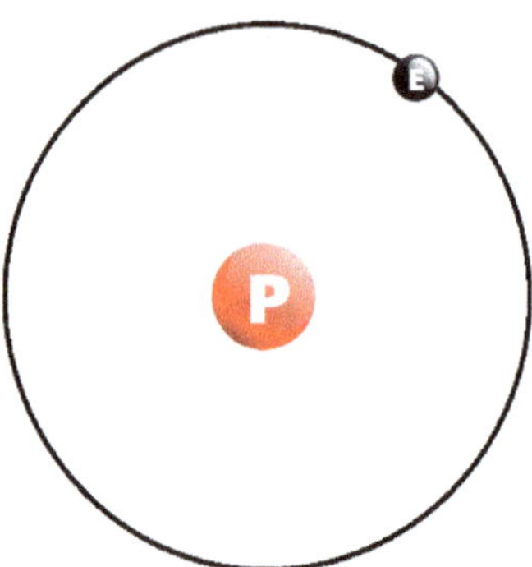

- How can this happen?

Any system that absorbs energy tends to gain speed and increase in volume. This also happens with the atom.

If this phenomenon occurs with the Neutron, this could enable an increase in the speed of the particle that will become the Electron.

This process can occur with the increase in speed inside the Neutron, generating a greater Centrifugal Force in this particle, enough for it to break the internal cohesive forces of the building that makes up the Neutron.

At first glance, this seems like an absurd proposal, but after studying the Isotopes Deuterium and Tritium, the pieces fit together.

It would be a comparison between the atom and a vegetable seed.

This parallel makes it clear that the atom is born, grows and dies – a natural process, beyond nuclear fusion.

Natural chemical elements also have some analogies to the development of natural processes. Such as the expansion and growth of childhood, the power of youth, the balance and moderation of middle age and the stagnation of old age. This process occurs in objects, in the solar system, galaxies and in the universe. There is birth, growth and death – death, of course, as well as the other processes, is transitory and does not mean the extinction of the substance.

- Why can't the same process happen in the atom?

Adding pieces to the puzzle makes the viability of the theory even clearer.

In this figure, we have Hydrogen, Deuterium, and Tritium.

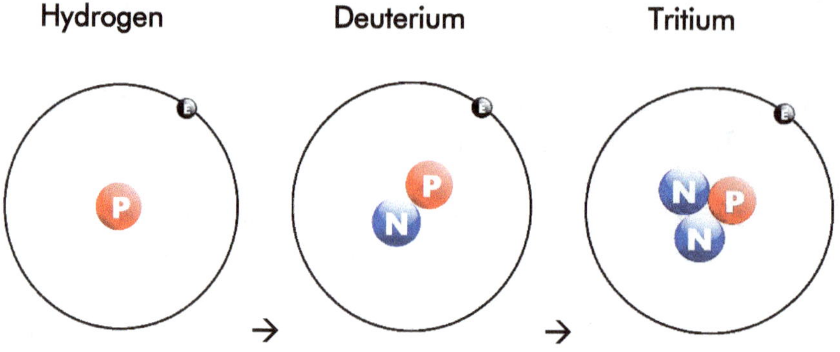

Each of the three has, evidently, just one Proton. The number of Neutrons, however, varies: none in Hydrogen, one in Deuterium, and two in Tritium.

It can be deduced, in analogy to what has been proposed so far, that Tritium has the potential to release an Electron from one of its Neutrons, given that, with the low neutrality of the system, there will be greater balance if the number of Electrons matches that of Protons.

Tritium is practically a potential Helium-3, requiring only the emission of an Electron by one of the Neutrons. See in the picture below:

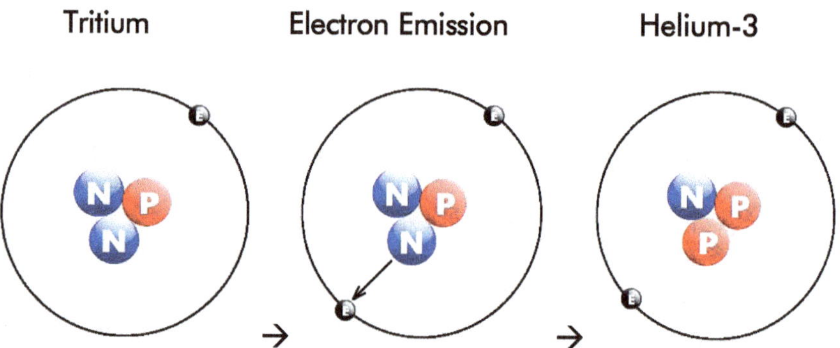

The possibility of a nucleus generating or capturing a Neutron allows physics and chemistry to reconstruct their theories of the development of the atom, of its growth from the inside out.

Once a new Neutron is generated in this Helium-3, it will be able to balance itself as the element Helium of the periodic table.

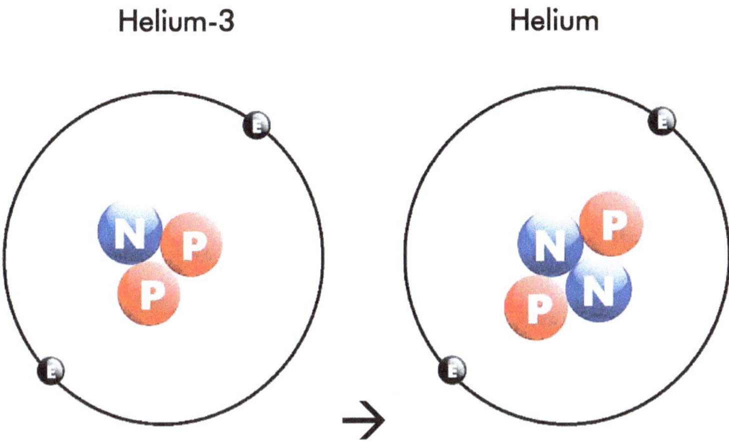

Transmutation from Hydrogen to Helium:

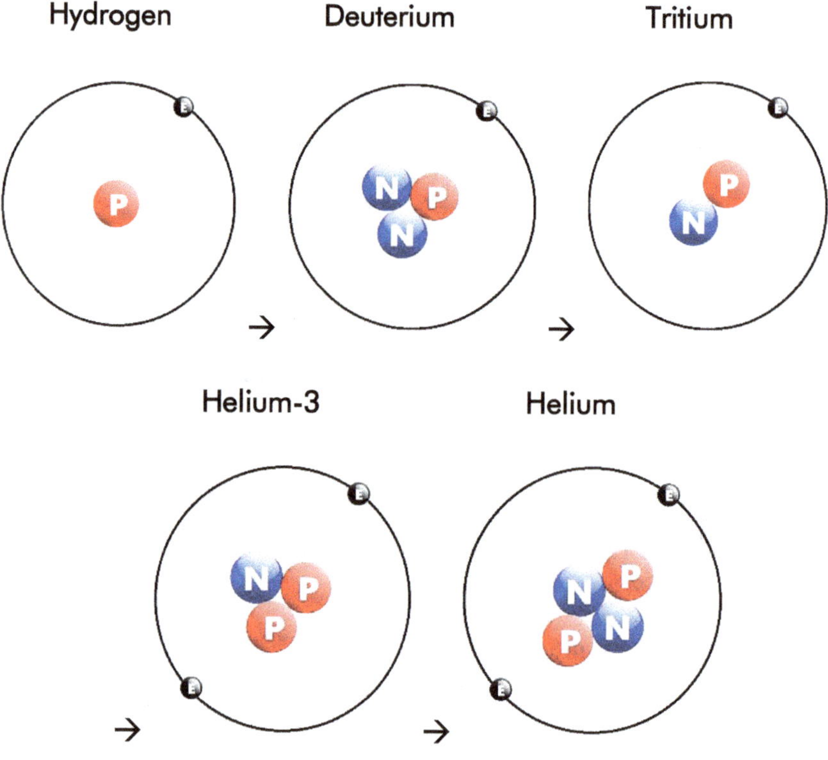

These proposals are simplified for easier understanding. Naturally, there are many questions, such as:

- How do Neutrons arise in this process?

- How long does this process take?

There is important evidence of the possibility of the Neutron emitting an electron: the masses of the Neutron, the Proton and the Electron. It will just require observing.

Particle	Mass	Load
Neutron	$1,675 \times 10^{-27}$ kg	Zero
Proton	$1,673 \times 10^{-27}$ kg	$1,6 \times 10^{-19}$ C
Electron	$9,109 \times 10^{-31}$ kg	$1,6 \times 10^{-19}$ C

(Mass in Kg)
Proton =
0,000000000000000000000000001673 (A)

Electron =
0,0000000000000000000000000000009109 (B)

Sum =
0,0000000000000000000000000016739109 (A + B = D)

Neutron =
0,000000000000000000000000001675 (C)

Mass of Neutron C – (sum of mass of Proton A + mass of Electron B)0,000000000000000000000000001675 -
0,0000000000000000000000000016739109 =
0,0000000000000000000000000000010891 (mass difference E)

0,0000000000000000000000000000009109 (mass of Electron B)

0,0000000000000000000000000000001782 (mass E – mass B = mass F)

0,0000000000000000000000000000010891 /
0,0000000000000000000000000000009109 =
1,19563 (G) = (mass E / mass B) = Factor between mass E and mass B)

<u>Note:</u>

- The mass of the Electron is on the same scale as the difference between the masses of the Neutron and the Proton.
- The difference between the masses of the Neutron and Proton is equivalent to a little more than the mass of two Electrons.
- When subtracting the mass of the Electron from the difference between the masses of the Proton and the Neutron, the equivalent of approximately 120% of the mass of the Electron remains.
- The resulting mass, subtracting the three particles (E), is probably in the form of an energy field, or possible loss of part of this mass to the external environment of the nuclear system.
- The mass (E) can explain what binds two Electrons when Quantum Entanglement occurs.

Note that the Neutron has a slightly larger mass than the Proton and that the Electron's mass is much smaller than that of the Neutron and Proton.

The hypothesis is that the small difference between the masses of the Proton and Neutron can justify the transition.

The hypothesis is that what remains of the difference between a Neutron and a Proton, after subtracting an Electron, is part of an energy cloud that connects Protons and Electrons, permeating them, and, perhaps, the loss of another particle.

This hypothesis is also based on the possibility of the binding energies acting in the form of vortices.

By logical deduction, there is a mechanism that mediates between a (+) charge and a (-) charge, and the most likely natural process would be the rotational direction of the vortex.

Below is an illustration of the proposal for atomic reality:

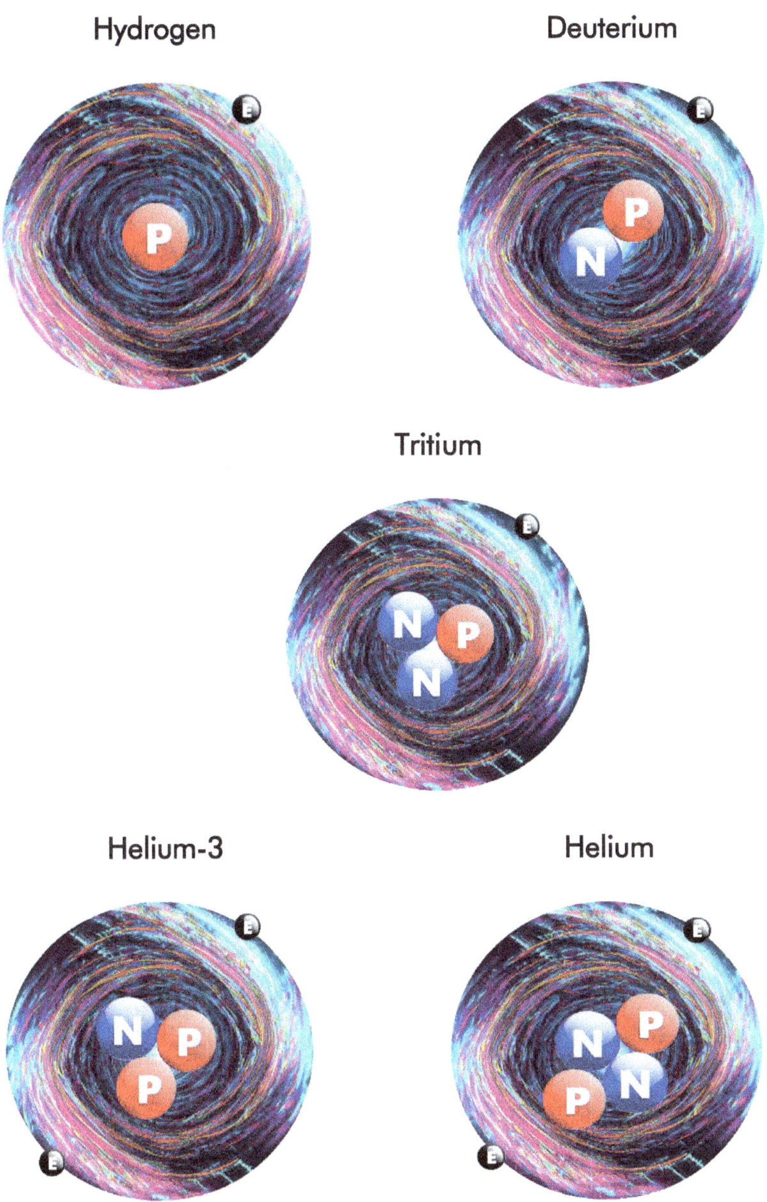

If the electron that will be emitted to form Hydrogen is contained in the Neutron, it is certainly orbiting the system of this composite particle. When it leaves the system, as proposed, it would logically do so in a spiral, following the sphere and, at the same time, moving away from the Neutron particle. Thus, the outward movement of this electron will form a small tornado. This vortex will be the starting point for the Neutron to become a Proton and the orbit of the electron will be its end. In the region between the nucleus and the electron, an electronic cloud will form, the link between the two.

This is the beginning of the atomic opening that will culminate with the element Uranium, where the Centrifugal Force of the electrons of the last layer will promote the escape of these electrons from the system due to the instability of the atomic building.

The atom is made up of a set of electronic vortices that balance each other and can complement each other in chemical reactions, forming more complex and functional structures.

The Neutron has two Down Quarks and one Up Quark, while the Proton has two Up Quarks and one Down Quark.

The hypothesis is that, when an electron leaves the Neutron, one of the Down Quarks changes positions to an Up Quark and, by logical deduction, this electron must be directly associated with the Quarks by the difference of charge.

Note:

It is known that in more complex and unstable nuclei, there is decay with the emission of electrons to give balance to the system, thus altering the Quarks.

In the specific case of the probable origin of Hydrogen from the Neutron, there is no decay, but a volumetric growth, the expansion of the planetary system.

This process would be similar to a star starting a planetary system with the first planet.

I want to report what preceded this information and conclusions about obtaining Helium without nuclear fusion.

In one of my books, I already warned about atomic growth without the need for nuclear fusion, but I did not see the more detailed path of how this could occur.

So, I imagined that the simplest path would be to start at the beginning of the Periodic Table, that is, with Hydrogen.

Another important factor was the information I discovered while researching the origin of Helium here on Earth.

The logic does not match the hypotheses of its formation.

Some interesting information about the formation of Helium here on Earth.

According to science, the formation of Helium Gas here on planet Earth occurred with the emission of Alpha particles (Helium nucleus) from the elements Uranium and Thorium for thousands of years. In this way, each Alpha particle (α) captured 2 electrons for the formation of the new element.

With this information, I did some research on this possibility, since significant quantities of Helium have already been extracted from the subsoil and many Uranium and Thorium mines have been discovered due to the use of these elements in the production of energy and for other purposes as well.

Helium is usually extracted together with Natural Gas. So, it was just a matter of comparing the volume produced and the geographical points of extraction. In other words, according to the theory in vogue, where there is Helium there is Uranium and Thorium, given the relationship between the generating source and the element generated.

When I started the comparison, the information did not match the theoretical data. So, I present what I found in a simple and even informal way. I do the calculations in a way that is generous to the theory in vogue.

The sources, found on the internet, are given credit.

World's Largest Producers of Uranium, Thorium and Helium

World Uranium Production (%) - 2019	World Thorium Reserves (million tonnes) - 2020	World Helium Production (mil. m^3) - 2020
Australia – 28	India – 1.070	USA - 74
Kazakhstan – 15	Brazil - 632	Qatar - 45
Canada – 9	Australia - 595	Algeria - 14
Russia – 8	USA - 595	Russia - 5
Namibia – 7	Egypt - 380	Australia - 4

Note:

The US produces 1% of the world's uranium, but is the world's largest producer of helium.

Qatar is the world's second-largest producer of helium, but does not produce uranium.

Source Credits:

Uranium Reference: World Nuclear Association

Thorium Reference: en.wikipedia.com

Helium Reference: www.statista.com

Calculations of Helium Production in the World

World production of Helium in m^3 in 2020.

140.000.000 (140 Millon m^3).

If this production remains the same for the next 30 years (estimated production in Qatar), we will have the following Helium production:

(140,000,000 x 30) = 4,200,000,000 (4.2 Billion m^3).

Although it is a hypothesis, it may be a higher or even lower value. Even with these variables, it is still a considerable value.

Density of Helium = 0.1785 kg/m^3.

Therefore, (4,200,000,000 m^3 x 0.1785 kg/m^3) = 749,700,000 kg

For tons ---> (749,700,000 / 1,000) = 749,700 tons in 30 years of production.

This is only an estimate for 30 years of global production of Helium from Natural Gas. It does not take into account what has already been produced since its discovery, what has already been lost in the various forms and new sources yet to be discovered.

Currently, companies are carrying out promising studies for new sources of Helium, and the global volume of extraction will certainly increase.

In the decay of 238U to 234U, an α particle is released that generates the same amount of Helium. In this way, the decay of 234U to 230Th releases another particle. Therefore, with these 2 decays, 2 atoms of Helium will be formed. Logically, from 1 atom of Uranium, 2 atoms of Helium will be produced.

The (α) particle is equal to the nucleus of the He atom, and when this nucleus captures two electrons, it generates the Helium atom.

Mole of Uranium = 238 g

Mole of Helium = 4 g

If 238 g of Uranium can be used to generate 8 g of Helium, 238 tons of Uranium produce 8 tons of Helium.

If in 4.5 billion years only half (50%) decayed to generate the element Helium, then there should be twice the amount of Uranium where Helium originates.

We can generate a factor for a vision of increased Helium production.

238 / 8 = 29.75

If only half of the Uranium and Thorium produced Helium gas, we can say that this factor doubles, that is, for every 1 ton of Helium produced by radioactive decay, 59.5 tons of Uranium and Thorium were needed.

How many tons of Uranium will be needed to produce 749,700 tons of Helium?

If 238 ---------------------- 8

Y -------------------------- 749,700

Y x 8 = 238 x 749,700 ---------- Y = (238 x 749,700) / 8 ---□ Y = 22,303,575 tons of Uranium.

In 4.5 billion years, the amount of Uranium will have to double.

22,303,575 x 2 = 44,607,150 tons of Uranium.

Estimated production of Uranium and Thorium for 30 years.

Only for comparison purposes with the estimated production of Helium.

Below are estimates of global uranium resources in 2019, according to the World Nuclear Association: 6,147,800 tons.

Global demand of 67,000 tU/year.

In the case of Thorium, according to the same World Nuclear Association, the estimates of global resources in 2020: 6,390,400 tons.

Adding the estimates of global resources for these two elements (Uranium and Thorium) we have the value: 6,147,800 tons of Uranium + 6,390,400 tons of Thorium = 12,538,200 tons.

Assuming that in 4.5 billion years it took 44,607,150 tons of Uranium to generate Helium, we can estimate how many tons of Helium could be formed according to the world estimates of Uranium and Thorium.

44,607,150 tons of Uranium--[] 749,700 tons of Helium

12,538,200 ----------------------------[] Y

Y = (12,538,200 x 749,700) / 44,607,150 -[]

Y = 210,726 tons of Helium.

A value far below reality.

Observations:

• Despite working with estimates through serious studies and by competent entities, these data do not represent reality, as it is not possible to map the full potential of the minerals (U and Th) under study and the gas (He) produced by these minerals.

- With these calculations and information provided by official bodies, doubts remain as to whether all Helium gas actually comes from these sources of Uranium and Thorium.

- So far, only mathematical hypotheses concerning the data obtained with the production of the three elements involved in the cycle, Uranium, Thorium, and Helium, have been dabbled.

- The parameters used were from data close to the calculations (2022), thus, past variables were not taken into account (production and losses of Helium as it is a gaseous element and lighter than air).

- The measured sources of Uranium and Thorium are mines and deposits close to the surface.

- The Helium produced comes from the extraction of Natural Gas (deep regions).

- There are extractions of Natural Gas that do not have the element Helium in their composition in significant quantities, but have large reserves of Uranium and Thorium (in Brazil, for example).

- Looking at maps of countries or regions that produce Helium, generally Uranium and Thorium mines are not close to Natural gas extraction points. In the United States, for example, the main Helium extraction region is between Kansas, Oklahoma and Texas, and the Uranium and Thorium mines are in the West, North and even East of the country. In many cases, countries that produce large amounts of Helium do not have reserves or mines of Uranium and Thorium (as in Qatar, for example).

- Although I know little about Geology and Nuclear Physics, I believe that, for the formation of Helium through atomic decay, Uranium and Thorium would have to be in favorable physical conditions: physical space and elements to donate electrons to the Alpha (α) particles.

- When researching the composition of volcanic lava (material resulting from deep regions of the planet), I did not find information about the presence of the elements Uranium and Thorium.

Unfortunately, I do not have information about the drillings to find Helium and the presence or absence of Uranium and Thorium in the extracted material.

Although not much is known about the deepest regions from which Natural Gas is extracted, by deduction, the possibility of a large presence of Uranium and Thorium, even in very remote times, is very small. It can be assumed, with what has been seen so far, that the sources of Helium on the planet probably had other origins and processes.

- There is doubt about which elements provided the electrons to bind to the Helium nuclei and the quantity of these elements.
- It is not known conclusively whether the Earth, in its 4.5 billion years, already had the element Uranium from the beginning – a subject that will be discussed later. The potential of the current hypothesis, that of Uranium and Thorium, does not seem great.

Based on the cited studies, I suggest three possibilities:

Possibility 1 – The hypothesis in vogue, in which Helium is produced through the atomic decay of Uranium forming Helium. Theory based on the presence of Helium observed by spectroscopy presenting a yellow spectral band. Also observed in solar eclipses.

Possibility 2 – Atomic transmutation (without nuclear fusion) or the growth of the atom through the emission of electrons from the atomic nucleus, as well as the growth of this nucleus, giving balance to the atom, as described in this work according to my hypotheses.

Possibility 3 – Gases trapped during the cooling of the planet, reinforcing the hypothesis of the formation of the planet inside the mother star, or in the case of Earth, inside the Sun. This information follows the same sources as the previous case and my hypotheses, which will be reported later.

Among the hypotheses presented, I defend the following order of probability:

Possibility 3 – Due to the high volume of Helium present in the formation of the planet, the hypothesis of cooling of the Earth's crust and factors such as the large presence of Hydrogen and Helium in the Sun. This is the most likely hypothesis for the presence of Helium on planet Earth. A clue to prove this hypothesis is the composition of Jupiter (75% by mass of Hydrogen and 24% by mass of Helium). By deduction, Jupiter also comes from inside the sun.

Possibility 1 – Due to the large reserves or abundance of the elements Uranium and Thorium on the planet. A rapid process that has already been confirmed by science.

Possibility 2 – Atomic transmutation without nuclear fusion, that can explain the large amount of Helium in nebulae, and also the presence of dust in these systems. This process may be the most abundant in the universe, but we do not know how long this process takes, nor the conditions of the physical environment here on Earth. Only science can verify this possibility.

An important piece of evidence to illustrate this hypothesis was the recent discovery of a starless galaxy.

Presented at the 243rd meeting of the American Astronomical Society, the discovery challenges the traditional notion of such systems.

J0613+52 was detected by chance by a radio telescope. Located 270 million light-years away, the gas-rich formation has likely remained largely unchanged during the 13.8 billion years since the Big Bang. An unexpected twist of fate has yielded an extraordinary scientific discovery: the Green Bank Radio Telescope (GBT) accidentally pointed to the wrong coordinates and found the galaxy J0613+52, says Karen O'Neil, chief scientist at the Green Bank Observatory. Located 270 million light-years from Earth, it is notable for its apparent lack of stars, manifesting itself as a haze of gas, typical of normal galaxies, but floating alone in space. What makes it unique is its structural similarity to known spiral galaxies, such as the Milky Way or Andromeda, but without the presence of visible stars.

More details later in the hypotheses about nebulae.

Note:

Regardless of whether possibility 2 is proven for academic and scientific purposes, this could be the beginning of an experiment for future production of Helium and Helium-3 for commercial purposes, energy production and research.

Companies should investigate and invest in this possibility, after all, the reserves of Helium on the planet are finite and worrying, scientists warn, and Helium is increasingly used in high technologies, such as aerospace, IT, medicine, instrumental research, semiconductors, and others.

Chapter

The Death of the Atom

In the relative world, everything that is born must die.

This cycle involves all phenomena, from the atom to life.

It is important to highlight, and has already been mentioned, that nothingness generates nothing. So, death is not the extinction of something, but merely a moment of transition of situation.

If energy was concentrated to form matter, that is, one of the dynamic conditions of energy that generates the sensation of solidity, it would be natural that, at some point, this process would return to the initial condition of manifestation.

According to the hypotheses presented in this work, from the moment the elementary particles were generated, the construction of matter began. At that point, there is no turning back, because, if it was born, it will develop and die.

Naturally, this process applies to all corners of the universe, governed as it is by the Laws of Physics and Chemistry. Therefore, at any time or place, everything repeats itself with precision.

Since the atom began its journey in the form of Hydrogen, it is known that, naturally, the last element is Uranium.

It is observable that elements are born and developed with the generation of electrons and the growth of the nucleus. This occurs in the growth phase, and, in the cases in the vicinity of Uranium, atomic instability begins to manifest itself.

The stable structure in the middle of the development cycle of matter tends to disintegrate due to the internal forces that promoted the opening of the atom, one of which is the Centrifugal Force.

The electrons in the last layer reach such high speeds that, because they are on the periphery of the atom, the forces that hold them to the nucleus are not capable of keeping these particles attached to the atomic body. In this way, the electrons leave, unbalancing the system, which in turn also promotes the loss of part of the atomic nucleus.

This is the process known as radioactive degradation. With this process, new elements are generated and energy is lost.

This is the process of the slow death of the atom and, as mentioned, the transition to other situations.

Transforming this concept of birth, growth and death of the atom into a geometric model, it would be an expansion vortex, that is, opening by centrifugal force.

It can be seen that the phenomenon of concentration followed by expansion generated a vortex of concentration on itself and of expansion up to a certain limit. We could compare it to the process of preparing a milkshake.

The milk and chocolate are in the blender. When the motor is turned on, the circular movement promotes centrifugal force throwing the mixture to the sides and it reaches a maximum height according to the energy used in the system and returns on itself by centripetal force.

This simple example is important to demonstrate that the two parts complement each other: the duality of the process forming a single body.

In the case of the cup of the home appliance, this is the limiting factor of the mixture. In the case of matter, the force created by energy clouds is the limiting factor.

We can draw another important comparative deduction: the occurrence of repulsion by centrifugal force being complemented by the attraction of centripetal force. The (+) and (-).

It is clear that the universe develops with these back-and-forth behaviors over itself. Always, however, at a different time or point.

By deduction, the volume (or magnitude) of these processes is linked to the potential of each system, from an atom to a galaxy.

I consider these factors to be extremely important for understanding the phenomena and laws that govern manifestations in the universe.

Remember that the curve is the vital path in manifestations.

To observe this type of vortex phenomenon more clearly, it is enough to study terrestrial cyclones.

They are formed by known variables, grow, and die due to wear and tear. However, the material that generated them does not cease to exist and returns to its previous form, wind, air in motion.

Measuring the lifetime of a chemical element, and even detecting it, is quite difficult. In addition to its complexity, the duration of its life and its transition may be beyond our ability to measure, and may even take millions of years, which leaves us in the dark.

There are several ways to measure the lifetime of an atom, but the vast majority are indirect or deductive methods.

Much of the energy that travels through the universe may have been matter at other times. These natural processes can also be carried out by human hands through fission or fusion, but instantly.

We can even compare this cycle of energy to matter, and the return to energy generating matter again, like a plastic material, which after forming an object, after its use, can be recycled, returning to an amorphous state and being modeled again as a new object.

Therefore, we can say that the death of matter is just a phase for its rebirth in another time and place.

Although we are talking about the quantum world, that is, about the atom, with a little observation, we can perceive this coming and going of energy conditions in our everyday world, which also happens in the macrocosm.

With these comparisons, one can already imagine or narrow down concepts between the micro and the macrocosm.

It is possible to start a thread in which phenomena are repeated in different dimensions (sizes).

The atom is worn down by other natural factors, such as the emission of waves from its body and light, for example. Thus, the atom already demonstrates its inner instability, since nothing is static in the universe.

If the atom receives a stronger order, it can change its structure to the extent of this interference.

In the quantum world, this happens in the most varied ways possible.

Human beings, when directly studying this latent world, can interfere through the presence and contact of research objects. This is because we are composed of atoms and have the power to interfere, even involuntarily, with these materials.

I believe that the energies that come from human beings can promote changes in nearby events, taking away their naturalness. This makes our vision of the phenomena of the quantum world disconcerting. Our interference can result in erroneous conclusions in a work.

Experiments should be isolated as much as possible, keeping them as close to a natural environment as possible.

The importance of this comment is for a possible comparison between quantum phenomena and cosmological phenomena.

52

Chapter IV

Antimatter

The birth and death of matter have been explained.

One of the great mysteries of physics is Antimatter, how it came about and where to find it.

The reasoning is the same as that for the formation of baryonic matter, formed by dynamic concentration of energy by a concentration vortex, that is, by Centripetal Force. With Antimatter the process could not be different.

What differs is the mechanism that gives the particle its characteristic charge, positive (+) or negative (-).

As previously reported on the subject of magnetism, the concentration vortex and the expansion vortex are mechanisms that generate the signals of the charges.

It was also mentioned that elementary particles rotate, maintaining a cloud of energy around them, and this is directly linked to the potential of this particle, its mass.

It has also been shown that the atomic nucleus, that is, the Neutron and the Proton, are structures similar to the atom. In other words, they are systems composed of several particles and it is the set of these smaller particles that generate the balance in these nuclear particles, that is, the Neutron and the Proton.

In the hypothesis presented about the birth of Hydrogen through the emission of an electron from within the Neutron, this event generated a change in this Neutron and it underwent mutation and loss of mass, transforming into a Proton. With this, the neutrality of this particle, with the loss of an electron with a negative charge (-), left the system with a positive charge (+), the proton.

So, we can deduce and raise the hypothesis that one (or more) particles within this proton are what maintain the positive charge (+).

Since science has already discovered the Positron, which is a particle with the same mass as the electron, but with a positive charge, this, in addition to proving the beautiful deductive work of Paul Dirac, may simply be a particle with a manifestation of a rotational vortex opposite to the electron, but normal in nature as matter.

As a clock needs gears that give balance and functionality to the set, the Positron can only be one of these pieces.

During a collision between two particles, I believe that one has a concentration vortex while the other has an expansion vortex. With the same mass, the Electron (-) and the Positron (+) will have a strong attraction, favoring their encounter.

The attraction due to the coupling of the complementary movements of the vortices of opposite charges, as demonstrated in the previous chapter on magnetism, may be the cause of the strong collisions between Electrons and Positrons – so strong, it must be said, that they generate explosions, resulting in the scattering of waves and light.

There are also forced collisions of particles in colliders. The former case, however, is natural.

This phenomenon strengthens the idea that matter is formed through the concentration of waves.

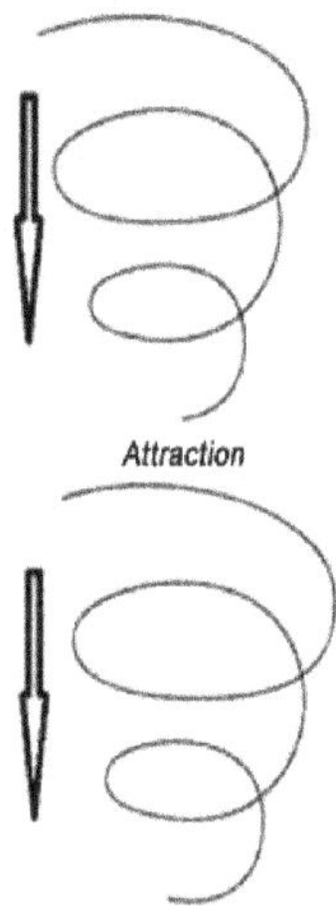

Figure 07 – Vortices of opposite charges attracting each other.

Because it is a subject treated in an erudite manner, or something exotic in the scientific community, the possibility of everything having its opposite in the case of particles and matter may have implications.

For example:

These particles may only be within sets, such as the Neutron and the Proton, due to the internal process of these composite particles.

This may explain the absence (or the small quantity) of this type of antimatter or free particles in the universe.

With the advancement of technology, and with the application of electromagnetic forces in the atomic system, it is possible to alter the intimate rotations of particles and even atoms.

If Positrons are particles with the same mass as Electrons, only having the opposite charge, and, in turn, are matter, then these particles suffer the attraction of gravity in the same way as Electrons.

In the course of this work, I will present the hypothesis of how Gravitation works. This information will make understanding the attraction of the Positron (Antimatter) easier.

In short, the formation of elementary particles with opposite charges is possible naturally due to the rotational process of the system. This process is linked to the concentration vortex and the expansion vortex. Remember the Milkshake.

For an Antiproton to exist naturally, for example, the internal system that makes up the proton would have to be constructed inversely with regard to the charges of the particles. I therefore believe that it is almost impossible, since everything in the universe follows a strict order and Laws.

Science can manipulate many things, but Nature will always obey the Laws and Order. And I always emphasize: every mechanism has to be the same in the past, as it is in the present, and will be in the future.

In short, science will not find what does not exist.

I do not speak as the owner of the truth, but according to the logic of the system.

Chapter V

Nebulae

Leaving the world of the atom and moving on to larger scales we find nebulae. They are very important for cosmology because they are sources of raw material. With the advent of modern telescopes, it has become possible to detect and investigate a variety of nebulae.

The vast majority of nebulae are already in a creative process, if we can say so, as they already contain stars, dust and other celestial bodies, in addition, of course, to the base gases, Hydrogen and Helium.

A nebula is actually a cradle of Hydrogen and Helium at first glance, but I want to insert a third component into this cradle, dust, that is, the chemical elements that make up this dust

Continuing with the hypothesis of this work, which is atomic growth by emission of electrons from the nucleus, it can be deduced that the first chemical element is Hydrogen and the second is Helium – a fact made explicit by the preponderance of the first in relation to the second.

The third component is the chemical elements above Helium, formed by atomic growth through the immeasurable passage of time. Their combination is dust.

Nebula A J0613+52, whose discovery was mentioned above, contained hydrogen, helium and dust, but did not generate stars or other celestial bodies. In light of this hypothesis, it can be assumed that this is a young nebula, formed recently.

There must be many nebulae like this in the universe, just as there are mature stars in which there are almost no more primordial elements in gaseous form, Hydrogen and Helium.

There are nebulae beginning their rotation process and generating their first stars. At this point, they are the seeds of future galaxies. As well as nebulae containing stars, but maintaining other shapes. In many cases, beautiful gaseous constructions.

The nebula is the clearest and most evident example of atomic growth as suggested here, because, despite the observer's short time here on Earth, nebulae appear in regions where mature structures are already found in the universe and, according to the explanation of the death of the atom, some structures disappear by atomic explosions or by simple transformation of matter returning to the energy phase.

In short, sporadic phenomena occur in the universe due to the dynamic concentration of energies in the form of vortices by centripetal force, forming atoms that make up a nebula and, after long cycles, these energies return through dynamic expansion by the centrifugal force of the matter (atoms) that make up the galaxies.

Some interesting observations for analysis:

- The presence of dust is observed in galaxies and nebulae, and this shows that this matter is grouped in these groups. However, if the dust were outside these clusters, the universe would be opaque and the lenses would see little in terms of the past and distances.
- This fact shows that the dust was probably generated in these clusters of matter.

- If a nebula does not contain stars, especially extremely old and unstable stars, there should be no dust in these nebulae.

- In the case of newly discovered galaxies close to what would be the beginning of the universe, there would also not be time for a star to be born, grow, age and explode, throwing matter in all directions.

- If an estimate were made of the number of stars that exploded and the amount of dust generated by this phenomenon, I believe that there is much more dust than the material generated by the explosions, due to the time of the cycle of these events.

- Another fact that must be taken into account is the preponderance of Hydrogen in relation to Helium and dust in nebulae, due to its implications.

- According to theories, dust is generated in nuclear fusions within stars, as this dust is composed of heavy chemical elements, that is, more complex than Hydrogen and Helium, and would take millions or billions of years to reach this point.

- Here, the idea is reinforced that the Laws of Physics and Chemistry must be the same now as they were billions of years ago. Therefore, the universe that we observe today must have the same characteristics as the primitive universe.

Chapter VI

Stars

- How to form a ball of hot gas in the universe?

This is an interesting mental exercise and clues have been given throughout the book.

The only process to unite amorphous gases at a point forming a sphere is by dynamic concentration of these gases by centripetal force in the form of a vortex.

What happens in the macrocosm is the same as what happens at the quantum level, forming elementary particles, as they obey the same dynamic principle.

- But what makes very cold gases transform into gigantic and extremely hot spherical objects?

The answer is simple, just observe a Quasar.

When the process of dynamic concentration of gases begins in a nebula, the concentration movement generates friction because these atoms are constrained by centripetal force, thus generating light, a lot of heat, and other forms of energy.

The heat from this friction and the movement will generate the appropriate conditions for Nuclear Fusion.

In this process of gas concentration, dust will also be dragged. Thus, it can be deduced that a star is born with heavy elements in its composition, probably in small percentages.

The spectrum of a star already shows its composition, that is, the predominant chemical elements in each phase of the stellar process.

The youngest stars are the brightest, clearest, and hottest and emit more ultraviolet waves and heat. As they age, the spectrum changes and the star loses brightness and emits waves of lower vibration.

Another natural process in stars is that their volume tends to decrease over time, and due to loss of energy, this process causes an increase in the density of these stars.

The star starts rich in Hydrogen and Helium; over time, these gases will transform into other gases and heavy metals.

One fact that supports the dynamic concentration proposal is the rotational movement of the star, because if it were a gravitational attraction type of concentration, the star probably would not have this type of movement, and perhaps would not have a spherical shape.

Going from a young star, where the simplest elements predominate, to an old star, where metals are almost all, it is possible to deduce that, in this last phase, the presence of radioactive elements must have increased. These conditions can generate instability in the star and, consequently, cause it to explode. Therefore, much of the aged matter returns to the energy phase and other solid parts travel through the universe.

Interesting evidence about the movement of some stars in the article below:

The Astrophysical Journal

A publishing partnership

Open access

The Internal Proper Motion Kinematics of NGC 346: Past Formation and Future Evolution

E. Sabbi1, P. Zeidler2, R. P. van der Marel1, A. Nota3, J. Anderson1, J. S. Gallagher4, D. J. Lennon5,6, L. J. Smith1, and M. Gennaro1

Published 2022 September 8 • © 2022. The Author(s). Published by the American Astronomical Society. The Astrophysical Journal, Volume 936, Number 2Citation E. Sabbi et al 2022 ApJ 936 135DOI 10.3847/1538-4357/ac8005

Article and author information

Abstract

We investigate the internal kinematics of the young star-forming region NGC 346 in the Small Magellanic Cloud (SMC). We used two epochs of deep F555W and F814W Hubble Space Telescope Advanced Camera for Surveys observations with an 11 yr baseline to determine proper motions and study the kinematics of different populations, as identified by their color–magnitude diagram and spatial distribution characteristics. The proper motion field of the young stars shows a complex structure with spatially coherent patterns. NGC 346's upper main sequence and pre-main sequence stars follow very similar motion patterns, with the outer parts of the cluster being characterized both by outflows and inflows. The proper motion field in the inner ~10 pc shows a combination of rotation and inflow, indicative of inspiraling motion. The rotation velocity in this regions peaks at ~3 km s−1, whereas the inflow velocity peaks at ~1 km s−1. Subclusters and massive young stellar objects in NGC 346 are found at the interface of significant changes in the coherence of the proper motion field. This suggests that turbulence is the main star formation driver in this region. Similar kinematics observed in the metal-poor NGC 346 and in the Milky Way's star-forming regions suggest that the differences in the cooling conditions due to different amounts of metallicity and dust density between the SMC and our galaxy are too small to alter significantly the process of star cluster assembly and growth. The main characteristics of our findings are consistent with various proposed star cluster formation models.

Figure – 08

In the Small Magellanic Cloud, a spiraling river of stars could reveal mysteries about the origin of the universe (Source: NASA/reproduction)

Article: The Astrophysical Journal:

doi.org/10.3847/1538-4357/ac8004 e doi.org/10.3847/1538-4357/ac8005

Chapter VII

Galaxies

Over the past two years, the James Webb (JWST) has been baffling science.

As observed through its powerful instruments, deep space appears to be similar to what is close to our time.

How can there be galaxies close to 13.8 billion years old with their mature structures?

Theoretically, there shouldn't even be galaxies in this period, but that's not what reality shows us.

This discovery requires new hypotheses about the construction of galaxies.

The hypothesis presented in this book is that the development processes of a particle or system occur through the same principle: the concentration and expansion of energy to form matter and, subsequently, the same process with matter to form larger structures.

It is important to highlight a characteristic of all processes:

Every process starts from a point and develops according to the potential of each system. In some more complex cases, where they are composed of matter, the mass can vary, but it will never be pulverized throughout the universe. There will always be a limit that will maintain the structure of this system.

We can give examples:
- An atom will have its maximum natural size.
- A tree will have its limit according to its species.
- Human beings will have their height limit.
- A galaxy will have its volume limited to its initial mass.

To make it easier to understand, we can imagine a situation where we have a mass of matter or energy to build something.

The result of the construction will always be the limit of the mass or the amount of energy available.

This rule explains why a galaxy does not break up in the universe, always maintaining a structure characteristic of its moment of evolution.

Like nebulae and stars, a galaxy will be the product of the progressive development of these structures. From the minimum, the maximum is built.

The first stars are formed in nebulae and this process always occurs dynamically and naturally.

When forming, the mass of a large celestial body can generate movements in the system around it, as well as magnetic fields, resulting from its energy fields and the circular movement originating from its genesis.

These movements are analogous to those of the microcosm: the vortex.

In a galaxy, an expanding vortex, its force will always go from the center to the periphery. At this starting point, the first celestial bodies will be found.

Since this process is the result of Centrifugal Force, naturally, these older bodies will tend to move toward the periphery of the galactic system in formation. When new stars are born, they will follow this same principle.

Remember that the Curve is the path that uses the least energy in nature and is used in developing natural processes.

As time goes by, the process of development unfolds: from the formation of the particle that will become the atomic nucleus, the atom, the nebulae of dust and gas, the stars and planets, and the galaxies. Everything happens following a natural path, transforming into increasingly complex constructions.

This entire phenomenon will have its limit in the Law of Compensation, just like in a terrestrial vortex. The limit of expansion will promote the return to the center of the galactic system in the form of concentration.

Reinforcing: if there is evolution in the systems, a star is born, grows, and dies; in this way, it returns to the energy phase.

If, in the case of a star, this process occurs due to internal instability due to its age, we can deduce that this star must be close to or on the outskirts of the galaxy. And, if the genesis of the system is in the center, the vast majority of the energy, after the explosion of a star, should return to it through centripetal force, as it creates a flow of concentration.

In this process, and science has already detected this, there is a compensation between the galactic vortex and its center (black hole); this is how the system is balanced.

Once again, it is demonstrated that there is a process that keeps the system cohesive, preventing its pulverization and disorder in the universe.

During the expansion of the galaxy, the stars become denser as they age, as metals replace gases. In this way, centrifugal force helps to move these denser massive bodies to the periphery.

I invite the reader and researcher to do, once again, a mental exercise: if the processes did not follow this model of concentration and expansion by vortices, wouldn't galaxies, for example, have already been pulverized, leaving the universe amorphous and long-distance observations precluded due to particles?

This principle of development of nature contains an order and follows this path throughout all coming into being.

In order to strengthen the hypothesis presented, we have recent evidence in the most distant galaxies detected by JWST. This, for now, because I believe that, soon, new discoveries will surpass this record.

The two record-breaking galaxies are called JADES-GS-z14-0 and JADES-GS-z14-1, the first being the most distant.

These galaxies already contained stars, gases, and, mainly, dust that, according to scientific hypotheses, would only result from the explosions of old and unstable stars.

For this to occur, the estimated time would be beyond billions of years.

As early as 2022, my book "Reconstructing the Universe" predicted that JWST would detect galaxies beyond 13.8 billion years ago – the estimated age of the universe. At the time, JWST was not yet in operation in space.

The new discoveries confirm the hypothesis, which predates them.

I reaffirm that, if the JWST has the capacity, it could exceed 13.8 billion years. If the JWST does not have it, another more sophisticated instrument will.

With superior technological conditions, quantities of mature galaxies similar to those close to the Milky Way will be observed.

This is not a prophecy or guesswork, but a logical deduction based on the principles presented in this work.

It is important to emphasize that the universe is far beyond our theories and hypotheses when it comes to lifespan.

In short: The idea of a time span for the universe is beyond our current intellectual capacity.

These 13.8 billion years of observed universe may be just a small phase on the scale of infinity.

The work of Dr. Roger Penrose, Nobel Prize winner in physics, is the closest to reality, without, however, following the basis of the Big Bang theory.

I believe that, with the rapid advancement of technology, these new ideas may have greater traction.

Chapter VIII

Galaxy System

Expanding the scope of studies on galaxies, we arrive at systems composed of galaxies, following an ordered rhythm in the universe.

Initially, science could only see, through its telescopes, the movement of these groups of stars moving away from each other.

As new instruments were placed on the ground and in space, new detections allowed for clearer and more time-consuming observations.

With the advancement in the use of computers in these observations, it was recently possible to simulate the order and path of these galactic movements.

Among these observations, I would like to highlight the movement of stars that move away through spiral movement, that is, a vortex, and gigantic filaments formed by galaxies, in which they move in the form of a vortex, heading towards large galactic clusters.

There are recent and very interesting studies carried out by the co-author of this work, Dr. Lior Shamir, on galactic behavior.

I will leave this information for Dr. Shamir to explain in depth through the facts and with his important conclusions.

In the same way that other processes develop through vortexes through centrifugal force, it would be no different with galaxy systems, because, as has already been demonstrated in several cases, everything follows a logic.

If the scope of observation increases even further, it can be assumed that there are even larger systems, composed of galaxy systems themselves, already beyond our current detection power.

At this point in the exposition, it is possible to conclude that quantum systems grow until they become the macrocosm itself using the same process: smaller vortices that compose larger vortices; a process of opening that is compensated by the return on itself by vortices of concentration.

If we could simplify the mechanism of the universe, the comparison of this dynamic would be the mixing of a Milkshake.

Figure 09 – Simplified illustration of the expansion of galaxy clusters.

In this figure, you can see the presence of galaxies rotating in a clockwise and counterclockwise direction.

I would like to draw your attention to the curved lines that extend from the observation point, which is planet Earth, to the galaxies. These lines represent the path of light in the universe over great distances.

Chapter IX

Behavior of Light in the Universe

Almost all the study of the universe occurs through waves in their variations and spectra.

In this case, I will focus on light waves.

Light travels at 299,792,458 m/s in a vacuum. It is a consensus that light travels in a straight line, except, of course, when it passes close to massive objects.

If massive objects were moved out of the way, would light travel in a straight line for great distances?

Our logic would say yes.

At this point, I will expand on the hypothesis we have been working within this book.

The vortex model is the key to the development of processes of all sizes, since, consuming less energy to obtain the greatest work, it is the easiest and most natural mechanism. Therefore, the curve is the best path for the movement of light.

We can imagine that light, like objects, must also obey this law of least effort. Therefore, when traveling astronomical distances, light can follow the curvature of the system. Until it reaches the observer, it must follow a curved path, in the form of a vortex.

Therefore, we have the illusion of seeing the universe expanding in a linear fashion, when, in fact, the observed waves do not arrive in a straight line, but through the curvature of the universe.

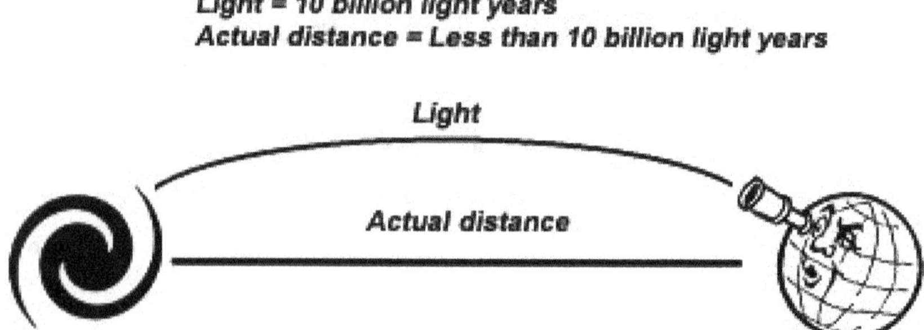

Figure 10 – The path of light; the real distance of the bodies.

This is a hypothetical view, if it were possible to mechanically and linearly measure the distance between celestial bodies.

Chapter X

Planets

Let's talk about the formation of our precious earthly home.

How are planets formed?

In the specific case of Earth, according to hypotheses, rocks that surrounded the Sun came together to form the planet, but it is worth remembering that the Laws of Physics are, by definition, conditions that are unchangeable over time. This is also the case with the gravitational force of rocks. The union of rocks by gravitational attraction forming a large group would therefore be an unlikely alternative, both in the distant past and in the present.

Highlighting that this set, this celestial body, rotates on its own axis and in harmony with its mother star and with other bodies in the solar system, including maintaining the spherical shape that characterizes the planets, moons, and stars.

Another question is raised by gas planets that, despite being spherical, rotate and orbit the Sun, would have to have a different formation process than rocky planets, such as Mars and Earth.

It is important to note that gas planets, such as Jupiter and Saturn, are much larger than rocky planets.

Comparing the atomic constitutions of Jupiter and the Sun, it is noted that they are similarly rich in Hydrogen and Helium.

The question remains:

- Why does the Solar System have rocky planets and gas planets?

According to the theory of the formation of the Earth, there should only be rocky planets.

Once again, I invite the reader to follow the reasoning.

This work develops the hypothesis of atomic growth by emission of electrons from its nucleus and has already discussed the formation of matter, atoms, stars, galaxies and systems of galaxies, even explaining galaxies with dust, but without old and unstable stars to explode and produce it. The same will be done to explain the formation of planets.

Since gas planets have similar constitutions to the stars they orbit, as Jupiter has to the Sun, it can be assumed that they were once part of the latter.

In parallel with the formation of Hydrogen, in which an electron is emitted from its nucleus and begins to orbit it, it can be imagined that a mass of gases, probably containing solid material, such as small concentrations of the dust that originated the star, was expelled from it, beginning to orbit it.

While still inside the star, these points of greater density behaved differently from their surroundings, ending up being expelled by the centrifugal force resulting from the star's rotation. Once propelled, they anchor themselves gravitationally to the parent star.

They would be like tiny, denser suns that would orbit around the sun.

Over time and with distance from the star, these masses tend to cool, losing volume and gaining density.

As seen with atomic growth, energy is the food of matter. Like a seed, matter tends to feed on the surrounding energy through speed, increasing the atomic mass, the nucleus and forming more electrons.

The result of this process will be the formation of heavier elements without nuclear fusion.

To illustrate in a simple way, the gases that make up the gas planets will transform into metals and in the future form the rocky body of the planet. With this process, there is an increase in density and a decrease in volume.

This process can be evidenced by the structure of the Earth, where the surface is more solid, because of the lower temperatures, and the center is more liquid, because of the higher temperatures – and, therefore, less dense.

This is a natural and logical process, in which any heated mass tends to balance with the colder external environment due to heat loss. In this process, cooling occurs from the outside in.

This resonates with the study of the element Helium, which observed that elements such as Uranium and Thorium are not found at great depths, as can be seen by their relative absence in the products of volcanic eruptions. The study suggests that the heavier elements must be closer to the Earth's surface.

These data suggest that rocky planets were initially gas planets and more voluminous. They may even have been the first to be emitted from within the Sun. In this case, volume could have been a determining factor, since smaller objects cool more quickly.

The (currently) gas planets, on the other hand, are larger and may be younger and, in the future, may become denser through the process of atomic growth. If this is the case, the Solar System will have more rocky planets.

The repeatability of processes in the growth and development of the universe can be perceived, always acting according to the same principles.

From the micro, it builds the macro.

What would a solar system be if not an atomic structure of gigantic size? And a galaxy if not a colossal molecular structure? Wouldn't the atomic nucleus, of protons and neutrons, be a tiny version surrounded by tiny satellites?

Scientists have already detected many planets outside the Solar System, for which this one serves as a parallel. This is yet another important piece of evidence that planets form within their parent stars.

This hypothesis facilitates the understanding of the structure and development processes of planets and stars – in addition to explaining the presence of Helium here on planet Earth. Incidentally, this topic was presented outside the logical expository sequence because more comparative details were needed to demonstrate the possibilities raised.

Chapter XI

Black Holes

Without a doubt, one of the subjects that most intrigues scientists is the Black Holes.

I would like to mention two types of natural movements that promote the development of phenomena in the universe.

The circular movement, which gives stability to systems, and the vortex movement (of concentration and expansion), in which centripetal and centrifugal forces manifest themselves.

- What does a black hole do?

It attracts energy and matter to its center. There is a consensus on this.

- But how does this process occur?

It is imagined or deduced that it is due to a strong Gravitational Force.

Here I try to present another possibility.

The process occurs by the dragging flows of energy and matter towards oneself through a manifestation similar to a terrestrial tornado.

Based on this work, I believe that the big question is:

- How does this phenomenon begin in the universe?

- What is the cause?

Here on Earth, we know why a tornado is formed, and we can deduce that in the case of a black hole, the principle may be similar.

Since many black holes are the center of galaxies, we can imagine that they play the role of dynamic concentration of the system, while the rotation causes the expansion of the galaxy.

With these two hypotheses, we return to the idea of the duality of systems, that is, a galaxy is formed by these two moments, in which one process complements the other, taking us back to the Milkshake example.

In a tornado, it is not gravity that attracts matter and gases to itself, but the typical movement of its process.

In the case of micro systems, or macro systems, such as a black hole, the process is the same, however, in the field of energy and matter. This principle does not prevent the presence of matter inside the black hole that generates gravity in the form of attraction as well.

With these hypotheses, we can imagine that, in a galaxy, everything tends to expand, and balances itself by returning the energy resulting from the dematerialization of matter to its center: the Black Hole.

If we could simplify this system, it would be a kind of cosmological drain.

As this book demonstrates in a very simple way, while trying to show the guiding line of phenomena, what exists is an eternal transformation in systems.

In the specific case of the black hole, it is just another important piece to provide balance and to form matter again through the constraint of energies, concentrating them again.

With this, the cycles extend to infinity, but always on an increasing scale.

Another fact to be highlighted is that a black hole is not a phenomenon without reference, because this type of phenomenon is nothing more than something manifesting itself on absurd scales in relation to us.

We could say that there are several black holes of different sizes, from the microcosm to the macrocosm.

Philosophically speaking, a black hole is not a cemetery of energies and matter, but a nursery of new manifestations. There one dies, and there one is reborn in new cycles.

Answering the two questions in this chapter, this natural phenomenon would begin, by deduction, in a process starting from the micro and growing to the macro.

In the case of the cause, I consider this question a mystery to be researched. One fact, however, is important to science:

This manifestation in the form of a vortex is the only model or process that promotes natural concentration and expansion in the universe.

This is because it consumes less work, that is, it is the most economical process in nature.

This is how nature is: perfect, consuming the minimum to produce maximum efficiency.

Regarding the possibility of black holes being formed after stellar explosions, I do not know if it is possible.

In this case, I cannot give an opinion.

Our time as observers is insignificant in relation to the great universal phenomena. Therefore, it is difficult to follow the processes.

80

Chapter XII

Gravitation

Gravitation, another of the great mysteries of science.

This subject was left for last on purpose so that the reader can have the critical mass to follow the reasoning on this topic.

In four moments, processes that generate the attraction of energies and matter to themselves were mentioned.

- The process of formation of the elementary particle.

- The hypothesis of the process of magnetism.

- The terrestrial tornado.

- The black hole.

In these four examples, the mechanism of action is practically the same, that is, there is a movement in the flow of energies or matter, which drags towards itself what is around it.

The force of attraction is directly linked to the potential of the phenomenon.

It is important to emphasize that this type of attraction process is part of two phenomena that complement each other: Expansion and Concentration. Therefore, we can study this possibility in the atom, that is, in the matter in which we live and of which we are constituted.

- How can the atom generate Gravitation?

The life of the atom was explained at the beginning of this book: it originates from the concentration of energy, stabilizes itself through the balance between circular motion and centripetal force, and opens itself through centrifugal force.

In short, the result of a cycle of two halves makes us imagine that, in reality, the atom behaves like a small vortex. As already described.

I want to emphasize this process of duality and stability of sets. Whether they are particles, atoms, molecules, tornadoes, galaxies, etc.

If this principle governs the phenomena and keeps them distinct, we can hypothesize the formation of gravity of matter.

When the atom reaches its final phase, the last natural element of the periodic table, Uranium, the deconstruction of this atomic edifice occurs. This process occurs with the loss of part of the nucleus and electrons that leave their orbit.

- In a natural process, what would be the direction in which these electrons escape?

- Linear or rotational?

Since the curve is the path of least work, the spiral waveform output is the most obvious possibility.

Another important observation is that these electrons on the atomic periphery have the highest speed and the tendency to open up like a vortex opens, naturally, due to the centrifugal force acting on the electrons.

Therefore, we can imagine that these electrons leave as a continuation of the model in which they were incorporated by attraction to the nucleus.

The systems do not pulverize, but generally remain close to their genetic center. The exit of these electrons must reach a point of equilibrium and return back to matter. After all, they were born from it.

As in the process of making a Milkshake, the expansion returns as a concentration, closing the natural process of duality.

These electrons probably form a similar process, or perhaps even an electron cloud at very high speed.

This return on itself allows the atom to attract energy and matter or even energies that enter its field of action.

Logically, for something to be attracted, it must have a connection with what attracts it.

It was found that mass attracts mass and that the greater the mass, the greater the attraction; this phenomenon is known as gravity.

From this, it can be deduced that gravity is not particles (Gravitons), but a phenomenon originating from matter itself.

With this fifth process explained, we can imagine that Gravitation is associated with the size of each system, from the micro to the macrocosm.

This leads us to imagine that the Gravitation of each system keeps sets of matter, such as solar systems, galaxies and sets of galaxies, together throughout the universe.

This Gravitation coordinates and controls the entire universal system.

Gravity works in a manner analogous to magnetism: its field opens into a vortex of expansion and, after equilibrium, returns on itself as a vortex of concentration, bringing the matter and energy that are under its field of action.

Gravity, on Earth, is linked to the magnetic field in relation to its mass.

Gravitational and magnetic processes work in parallel: their fields open into expansion vortices and, after equilibrium, return upon themselves as concentration vortices, bringing the matter and energy that are under their field of action.

In a critical analysis, only this model of manifestation can make the connection between two bodies or energies and pull them toward the generating source. This alignment of vortices is fundamental to understanding this gravitational mechanism.

In the case of the gravitation to which we are subjected, this manifestation may be above the speed of light, making it difficult to detect the process.

In summary, we have the following cases of attraction in the universe:

- The immediate duality of expansion and return through concentration to the point of origin, like Centrifugal Force returning with Centripetal Force = Gravitation.

- The power of dragging whatever is within the reach of the process through Centripetal Force, such as, for example, the Black Hole and the Earth's tornado.

- Magnetic forces through the alignment of positive charges with negative charges.

Another hypothesis is that the phenomena will have their manifestations at similar speeds, whether in the microcosm or in the macrocosm.

It may seem that, due to its gigantic size, what would take an infinitesimal amount of time to happen in an atom or particle would take thousands of years to happen in a galaxy. However, the time would be the same, whether the energy is returning from the outskirts of the galaxy to its black hole (its center) or whether it is returning from the electron to the nucleus of protons and neutrons of an atom.

Second Part

Lior Shamir

86

Chapter **XIII**

Empirical evidence of bias in the redshift model and its implications on cosmology

Since the 1920s, the Big Bang theory has been the most dominant cosmological theory. It was proposed by George Lemaitre and became well known through the work of Edwin Hubble. The lading observation leading to the Big Bang theory is the consistent link between redshift of the galaxy and its distance from Earth. The farther the galaxy is from Earth, the higher its redshift, meaning that it is moving away from Earth at a faster velocity compared to galaxies that are closer to Earth. That suggests that all galaxies started their journey in space from the same location at the same time, and therefore in a Big Bang.

Although the Big Bang theory is still the leading theory that explains the nature of the Universe, several recent observations have been challenging the Big Bang theory. One of these observations is the *Hubble constant (Ho) tension*. The Ho tension means that the expansion rate of the Universe as observed by the velocity of galaxies is not the same as the expansion rate of the Universe as observed by the cosmic microwave background radiation. Because there is only one Universe, one of these measurements is expected to be incorrect, but it is unclear which one, as both were teste and are assumed to be correct.

Another observation that challenges the Big Bang theory is the discovery of large and mature galaxies observed by James Webb Space Telescope (JWST) at very high redshifts. For instance, the galaxy

JADES-GS-z14-0 is estimated to have a redshift of 14.2, which means that the galaxy the we see from Earth is less than 300 million years after the Big Bang. The problem is that a large and massive galaxy is supposed to be much older than 300 million years. Therefore, the existence of such galaxies challenges the Big Bang theory.

Figure 1. JADES-GS-z14-0 is a galaxy with estimated redshift of 14.2. That means it is less than 300 million years after the Big Bang. The existence of such galaxies challenges the Big Bang theory since large massive galaxies are not expected to form in such a short time. Image credit: Credit: NASA, ESA, CSA, STScI, B. Robertson (UC Santa Cruz), B. Johnson (CfA), S. Tacchella (Cambridge), P. Cargile (CfA).

In fact, large and massive galaxies at very high redshifts were also known to exist before JWST was launched[1]. These puzzling

---

[1] Neeleman, M.; Prochaska, J.X.; Kanekar, N.; Rafelski, M. A cold, massive, rotating disk galaxy 1.5 billion years after the Big Bang. *Nature* 2020, 581, 269–272.

observations are in conflict with the Big Bang theory and the LCDM model, and require some modification. According the LCDM cosmology, these galaxies were not supposed to exist, as was predicted before JWST was launched[2].

The fact that these galaxies exist, and given our understanding of the nature of galaxies, can have two explanations. Either the LCDM model is not correct, or the redshift model is not correct. But it is clear that the two cannot co-exist without modifications. But if the redshift model that is used today is not correct, it might have implications on many other theories that are determined by the analysis of the redshift, an that includes the Big Bang theory.

One of the assumptions used by the redshift model is that the rotational velocity of the Earth within the Milky Way, and the rotational velocity of the observed galaxies, is negligible. The Earth rotates around the center of the galaxy in velocity of $\sim$220 km/sec, and the observed galaxies rotate in similar velocity. That velocity is obviously very small compared to the linear velocity of the galaxies as they move away from the Milky Way, which can be thousands or tens of thousands of kilometers per second. Because the rotational velocity is so small compared to the linear velocity, most redshift models ignore the rotational velocity and considered it as negligible.

The problem with that assumption is that the physics of galaxy rotation is not understood. Early astronomers assumed that the physics of galaxy rotation was driven by Newtonian physics. That assumption has been challenged for several decades, until Vera C. Rubin showed that the physics of galaxy rotation does not behave in a manner that was expected by the known physics. Stars that are close to the center of galaxies rotate in the same velocity as stars located in the outskirts of the galaxy, which is in violation of Newtonian and Keplerian physics. The leading explanations to that provocative observation is dark matter and modified Newtonian dynamics (MOND). According to the

---

[2] Cowley, W.I.; Baugh, C.M.; Cole, S.; Frenk, C.S.; Lacey, C.G. Predictions for deep galaxy surveys with JWST from ΛCDM. *Mon. Not. R. Astron. Soc.* 2018, *474*, 2352–2372

dark matter theory the vast majority of the mass of the galaxies is invisible, and does not interact with light or with any other radiation. According to MOND, galaxies are driven by physics that is not the Newtonian physics. Other theories have also been proposed, but until now no theory has been proven. Ark matter is considered the most likely explanation, but its existence has not been proven, and it is still unclear why and how galaxies rotate. That means that when the redshift model assumes that the effect of the rotational velocity of galaxies is known and negligible, the redshift model is based on unproven assumptions about physics that is already known to be mysterious. That means that it is possible that the redshift model used in cosmology is not complete.

Preliminary signs to the unexpected effect of the rotational velocity of the rotational velocity of the galaxy can be observed in JWST deep field images taken in close proximity to the Galactic pole. The following image is a deep field image taken by JWST inside the field of Hubble Ultra Deep Field. The image shows a far higher number of galaxies that rotate in the opposite direction relative to the Milky Way compared to galaxies that rotate in the same direction relative to the Milky Way[3].

---

[3] Shamir, L., Galaxy spin direction asymmetry in JWST deep fields, *Publications of the Astronomical Society of Australia*, 41, e038, 2024

Figure 2. Deep field image taken by James Webb Space Telescope inside the field of HST UDF, and close to the Southern galactic pole. The image shows a much higher number of galaxies that rotate in the opposite direction relative to the Milky Way.

The analysis of the galaxies was done using a symmetric algorithm that can provide an objective analysis of the shapes of the galaxy arms. But even a simple manual inspection would show the same difference. The image has 24 galaxies rotating in opposite direction relative to the Milky Way (clockwise), and just 10 galaxies rotating in the same direction relative to the Milky Way. That is a difference of 140%. The probability of such distribution to happen by mere chance is ~0.012.

While the difference is obvious, the anomaly is very difficult to explain using the current cosmological models. In a homogenous universe as expected according to the Cosmological Principle, the number of galaxies rotating in opposite directions is expected to be the same, within statistical error. But JWST shows something completely different.

The difference is in agreement with Earth-based data that was released before JWST was launched. The flowing figure shows an analysis of ~1.3 million galaxies imaged by the DESI Legacy Survey. The analysis shows the differences between the number of galaxies rotating clockwise and galaxies rotating counterclockwise. The large number of galaxies allows to analyze the difference in the number of galaxies around each point in the sky. The sky projection shows that around the Southern Galactic pole, and in close proximity to HST Ultra Deep Field, there should be a higher number of galaxies rotating clockwise wise compared to galaxies rotating counterclockwise. This prediction was released in 2022[4], and before JWST was launched.

––––––––––––––––––––––––––––––––

[4] Shamir, L., Analysis of spin directions of galaxies in the DESI Legacy Survey, *Monthly Notices of the Royal Astronomical Society*, 516(2), 2281-2291, 2022

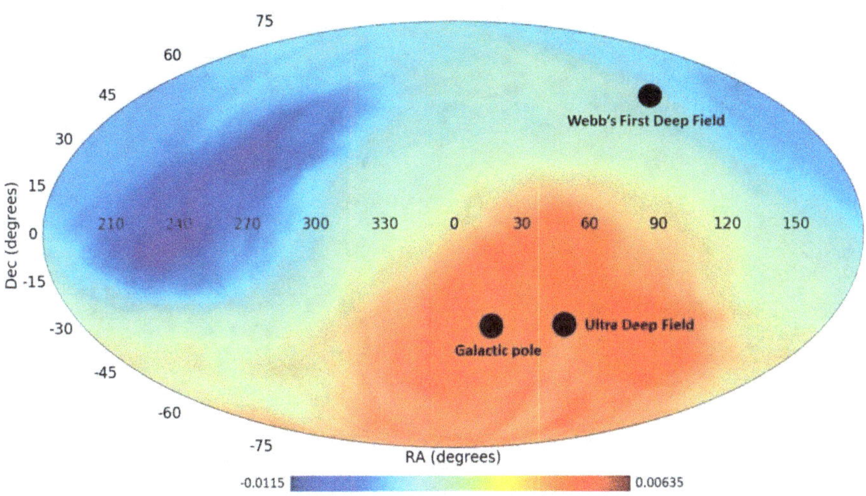

Figure 3. Analysis of direction of rotation of 1.3 million galaxies imaged by DESI Legacy Survey. The analysis shows that a higher number of galaxies rotating clockwise is expected around the Southern Galactic pole. JWST deep field image taken around that part of the sky showed exactly that.

The all-sky projection also shows the location of Webb's First Deep Field (SMACS J0723.3-7327). Webb First Deep Field is located in a part of the sky where no substantial differences between galaxies that rotate in opposite direction is expected. When looking at Webb's First Deep Field, the number of galaxies rotating clockwise is almost the same as the number of galaxies that rotate counterclockwise.

In fact, it is not merely the DESI Legacy Survey and JWST that shows that the number of galaxies that rotate in one direction is not equal to the number of galaxies that rotate in the opposite direction. All major digital sky surveys show the same observation. The number of galaxies that rotate clockwise around the Southern Galactic pole is higher than the number of galaxies that rotate counterclockwise. In the Northern Galactic pole the difference is inverse, and the number of galaxies rotating counterclockwise is excessive.

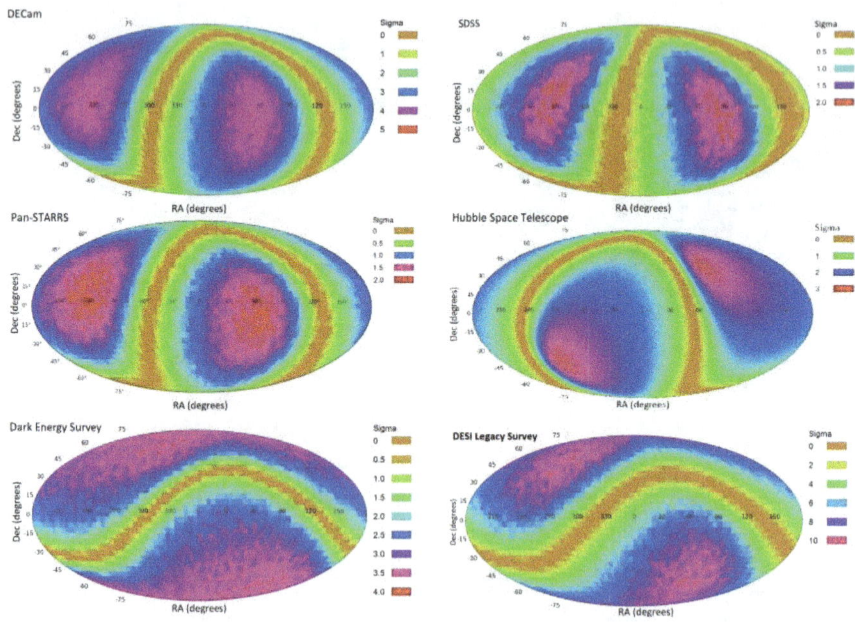

Figure 4. The statistical significance of a dipole axis formed by the distribution of the direction of rotation of galaxies. Telescopes that were tested include the Dark Energy Survey, the Sloan digital Sky Survey, Pan-STARRS, Hubble Space Telescope, DECam and the DESI Legacy Survey. All of them show that the spin directions of galaxies are not distributed randomly, and form a dipole axis.

The dipole axis formed by all of these telescopes and all experiments peak with close proximity to the galactic pole[5]. That might be a coincidence, but it is also highly possible that the rotational velocity of the Earth rotating around the Galactic pole is the cause of that anomaly.

---

[5] McAdam, D. and Shamir, L. (2023a). Asymmetry between galaxy apparent magnitudes shows a possible tension between physical properties of galaxies and their rotational velocity. *Symmetry*, 15(6).

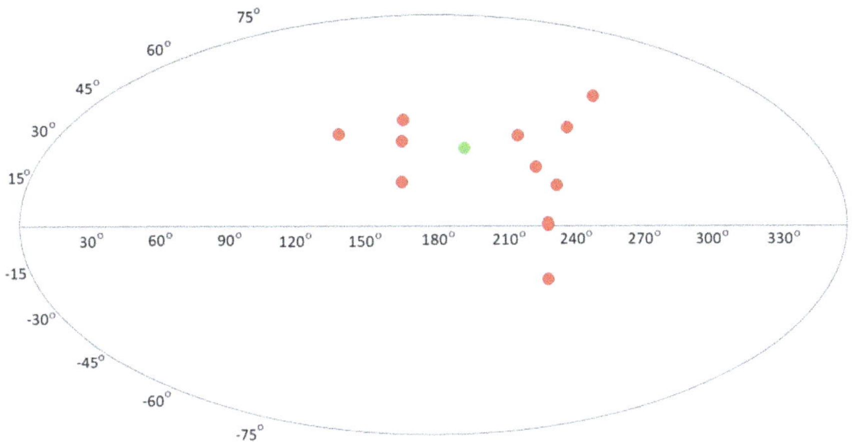

Figure 5. The peak of differences between the number of galaxies that rotate in opposite directions. All experiments show a peak with close proximity to the Galactic pole. While that can be a coincidence, it can also mean that the difference between the number of galaxies might be linked to the rotational velocity of the Earth around the Galactic pole.

The number of galaxies that rotate in one direction is expected to be the same as the number of galaxies that rotate in the opposite direction, within statistical error. Therefore, any statistically significant difference between the number of galaxies that rotate in opposite direction is unexpected. Because these differences are not expected but they certainly exist, there is something that is not considered in the existing cosmological models that causes it. The fact that the difference peaks around the Galactic pole suggests that the observation might be related to the rotational velocity of the Earth around the Galactic pole.

One possible explanation to that is that in the early universe galaxies tended to rotate in the same direction. In that case, the proximity to the Galactic pole is a coincidence. The other explanation is that the rotational velocity of the Earth around the center of the Milky Way relative to the rotational velocity of the observed galaxies leads to a change in their brightness as observed from Earth. The change in brightness leads to a change in their number as seen from Earth. That is, the number of galaxies that rotate in the opposite direction relative to the Milky Way is not really higher than galaxies that rotate in the opposite direction relative to the Milky Way. But because these galaxies are brighter, more of them are seen.

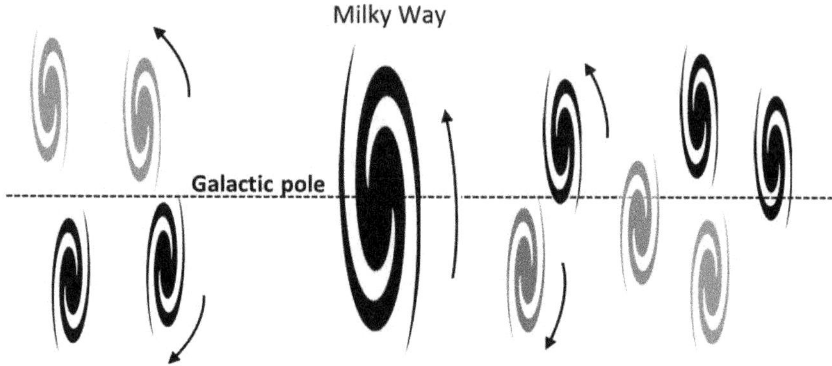

Figure 6. Galaxies that rotate in the opposite direction relative to the Milky Way are brighter than galaxies that rotate in the same direction relative to the Milky Way. Therefore, more galaxies that rotate in the opposite direction relative to the Milky Way will be observed.

The contention that a galaxy that rotate in the opposite direction relative to the Milky Way is brighter than a galaxy that rotate sin the same direction relative to the Milky Way is not unexpected. Due to the Doppler shift effect, galaxies that rotate in the opposite direction relative to the Milky Way are indeed expected to be slightly brighter than galaxies that rotate in the opposite direction relative to the Milky Way. That difference can be formulated by

$$F = F_0\left(1 + 4 \cdot \frac{V_r}{c}\right)$$

Where F is the observed flux, Fo is the flux of the observed galaxy if it is stationary relative to the Earth, Vr is the rotational velocity of the galaxy relative to the rotational velocity of the Earth, and c is the speed of light.

Assuming rotational velocity of the Earth around the Milky Way of 220 km/sec, the expected difference in magnitude is 0.006. That difference seems negligible. But because the physics of galaxy rotation, and especially its rotational velocity, is still mysterious and not understood to science, that equation might provide a difference in the flux that is much greater than the expected difference.

But when testing it empirically, the difference in the magnitude is far greater than 0.006. The following table shows the differences in the brightness (magnitude) in different bands between galaxies that rotate in the same direction relative to the Milky Way an in the opposite direction relative to the Milky Way. The analysis was done with $\sim 1.2 \times 10^5$ galaxies from the Sloan Digital Sky Survey, but the same difference is observed with any other relevant telescope[6]. Obviously, the comparison cannot be made using just a single pair of galaxies, since the Fo is not known. But when averaging the magnitude of thousands of galaxies, the average magnitude of each group of galaxies can be compared.

Band	Magnitude same direction	Magnitude opposite direction	dMag	P (t-test)
G	17.7095 ± 0.005	17.6948 ± 0.005	0.0147	0.0376
R	16.9893 ± 0.004	17.6948 ± 0.004	0.0148	0.0089
Z	16.4564 ± 0.004	16.4393 ± 0.004	0.0171	0.0025

The magnitude of galaxies that rotate in the same direction relative to the Milky Way compared to the magnitude of galaxies that rotate in the opposite direction relative to the Milky Way. Galaxies that rotate in the opposite direction relative to the Milky Way are, on average, brighter. The analysis is based on about 1.2×10^5 galaxies.

---

[6] McAdam, D., Shamir, L., Asymmetry between galaxy apparent magnitudes shows a possible tension between physical properties of galaxies and their rotational velocity, *Symmetry*, 15(6), 1190, 2023.

The same observation can be made with over 5,000 galaxies images by Hubble Space Telescope.

Band	Magnitude same direction	Magnitude opposite direction	dMag	P (t-test)
G	23.131±0.019	23.077±0.019	0.054	0.023
R	22.266±0.019	22.218±0.02	0.048	0.045
Z	21.358±0.017	21.323±0.018	0.035	0.087

Analysis of galaxies image by Hubble Space Telescope and rotate in the same direction relative to the Milky Way and galaxies that rotate in the opposite direction relative to the Milky Way. HST also shows that galaxies that rotate in the opposite direction relative to the Milky Way are, on average, brighter.

The fact that galaxies that rotate in the opposite direction relative to the Milky Way are much brighter than galaxies that rotate in the opposite direction relative to the Milky Way shows that the rotational velocity of Earth as it rotates around the center of the Milky Way galaxy affects the light emitted by these galaxies and observed from Earth much more than expected.

That indicates that ignoring the rotational velocity of the Earth can affect also the redshift of the galaxies. Because the redshift models ignore the rotational velocity of the Earth, the redshift of the observed galaxies might not be accurate, as it ignores a component that can affect the redshift much more than presently expected. Luckily, this can be tested empirically by comparing the average redshift of galaxies that rotate in the same direction relative to the Milky Way and galaxies that rotate in the opposite direction relative to the Milky Way.

The following table shows the average redshift (z) of galaxies that rotate in the same direction relative to the Milky Way and galaxies that rotate in the opposite direction relative to the Milky Way. The average redshift of these two groups of galaxies should be nearly identical.

Instead, it is clear that the redshift of galaxies that rotate in the same direction relative to the Milky Way is, on average, significantly different from the average redshift of galaxies that rotate in the opposite direction relative to the Milky Way.

Survey	Pole Hemisphere	Field size (°)	Annotation	# MW	# OMW	z_{mw}	z_{omw}	Δz	t-test p
SDSS	North	10×10	Ganalyzer	204	202	0.0996 ± 0.0036	0.08774 ± 0.0036	0.01185 ± 0.005	0.01
SDSS	North	20×20	Ganalyzer	817	825	0.09545 ± 0.0017	0.08895 ± 0.0016	0.0065 ± 0.0023	0.0029
SDSS	North	20×20	Galaxy Zoo	154	135	0.07384 ± 0.004	0.06829 ± 0.0035	0.0056 ± 0.0053	0.15
SDSS	North	10×10	SpArcFiRe	710	732	0.07197 ± 0.0015	0.06234 ± 0.0014	0.00963 ± 0.002	<0.0001
SDSS	North	10×10	SpArcFiRe Mirrored	728	709	0.06375 ± 0.0014	0.07191 ± 0.0014	-0.00816 ± 0.002	<0.0001
SDSS	North	20×20	SpArcFiRe	2903	2976	0.07285 ± 0.0007	0.071164 ± 0.0007	0.001686 ± 0.0009	0.04
SDSS	North	20×20	SpArcFiRe Mirrored	3003	2914	0.07113 ± 0.0007	0.07271 ± 0.0007	-0.00158 ± 0.0009	0.05
DESI	South	10×10	Ganalyzer	414	376	0.1352 ± 0.0027	0.1270 ± 0.0025	0.0082 ± 0.0036	0.018
DESI	South	20×20	Ganalyzer	1702	1681	0.1317 ± 0.0013	0.1273 ± 0.0014	0.0044 ± 0.0018	0.008

Because the distance of galaxies from Earth is not expected to be related to their direction of rotation, it can be assumed that the difference in the redshift is a feature of the rotational velocity of the Earth within the Milky Way compared to the rotational velocity of the observed galaxies. That means that the rotational velocity of the Earth within the Milky Way affects the redshift of the galaxies being observed.

A very interesting observation is that the redshift difference grows when the galaxies are farther away from the Milky Way. That means that the redshift bias is small at the smaller redshifts, but increase for higher redshift ranges. The following figure shows how the differences in redshift grow when the redshift gets higher. The analysis is done on relatively low redshift of less than 0.3, because galaxies at higher redshift are difficult to observe from Earth with sufficient details that allow to determine their direction of rotation.

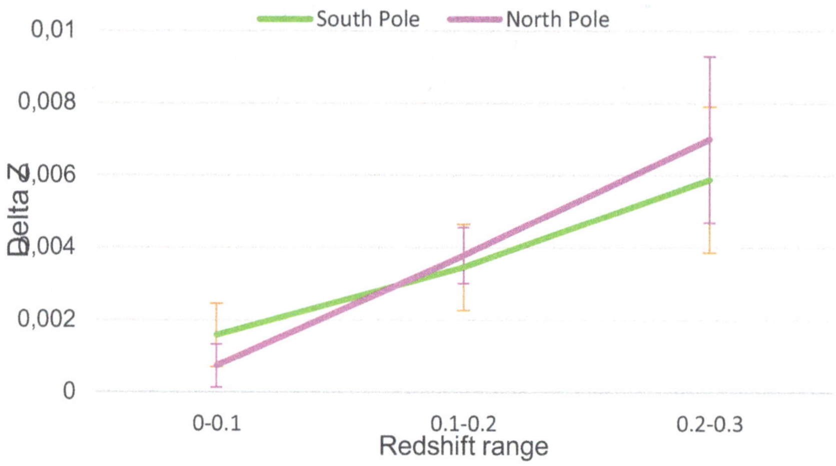

The difference in the average redshift of galaxies that rotate in the same direction relative to the Milky Way and in the opposite direction relative to the Milky Way. The analysis shows in both the Northern and Southern galactic pole that the difference grows with the redshift. That means that the redshift bias is higher at higher redshifts.

A higher redshift bias when the redshift is higher means that galaxies observed to have high redshift might have significantly different redshift than what is currently believed. That means that their real redshift is not nearly what their observed redshift is. That is due to the effect of the rotational velocity of the Earth relative to the observed galaxies. That can explain the mature and massive galaxies at very high redshifts. The galaxies do exist, but their redshift is different than what is currently being measured.

But the fact that the redshift bias changes when the galaxies are more distant from Earth can also have other meanings. The Big Bang theory is based on the correlation between the redshift of the galaxy and its distance from Earth. But if the redshift bias changes with the distance of the galaxy, the Big Bang assumptions might be partially driven by the bias in the measurement of the redshift rather than by the velocity of the galaxies. In that case more distant galaxies might have a different redshift, but that does not necessarily mean that they have different velocity.

Although such scenario is not likely, the bias in the redshift driven by the rotational velocity of the Earth within the Milky Way galaxy affects the measurements. That means that the age of the Universe is not the same age determined from the current analysis of the velocity of galaxies as they move away from Earth. The galaxies that we observe with very high redshifts are therefore probably far older than can be assume by using the redshift in the common model that ignores the rotational velocity of the Milky Way galaxy. Using the rotational velocity of the Milky Way in the redshift model will lead to a completely distance scales, and a new cosmology.

Chapter XIV

Conclusion

Our intellectual journey of this work has come to an end.

This is not about the end of subjects, just a pause, because scientific mysteries are eternal.

New concepts and ideas have been presented to the reader and researcher. This presented material is not the fruit of this author's imagination, but the result of more than 35 years of study, personal experiences, deductions, and information provided in the public domain.

There is no imposition of ideas here, but an invitation for the reader to penetrate the flow of this work.

The objective is to show the probable guiding principle of physical and chemical phenomena, the Laws, that govern the universe.

Certainly, there are Laws that we are unaware of and many phenomena may be under their control.

Knowledge is a bottomless well and those who think they have it all, even in their specialty, are deluded.

Today's certainties can be changed with new certainties or truths. Truth, in particular, has the depth of each mind's capacity.

It is a dive into the unknown, it is the infinity that reveals itself through each effort in progress.

I must warn you that only the shell of this universe that we detect, observe, measure and count has been explored.

There are universes that are intimately linked to this one, not because they are parallel, but because there is a continuity that we are still unaware of.

Here we report hypotheses and facts about energy and matter and we do not even touch on the spirit phase, as, for many, it may seem that it is dissociated from matter and energy.

We have not explored the religious factor, but I must warn you that, without this ingredient, the Formula of Everything does not add up.

What exists will always be part of the composition of the universe.

This first part began with the Movement, the basis of everything we know: without this dynamism, there would be no constructions.

In my opinion, the universal movement is <u>continually fed</u>.

A car does not continue forever with just one tank of fuel.

The primary types of movements for the grandiose universal construction were also determined, by hypothesis.

When the entire path from micro to macro is observed and logically analyzed, it is proven that it could not be any other way, that it could not be any other process. Try to imagine and you will see that things do not fit together in any other way.

With Mathematics, one can enter other paths, but it will not always be able to reflect the reality of possibilities and manifestations.

The movement in the form of a wave is what gives dynamism to energy.

The models of how to concentrate these dynamic manifestations, if we can describe them that way, are vital for the transformation of energy into matter, making the dynamism punctuate itself and, with that, the energy can be transformed into concrete building blocks.

This apparent chance has wisdom embedded in its behavior and this becomes repetitive in the form of a rigid, unshakable Law in its process.

The formation of points or, as we know them, elementary particles, is the beginning of the great construction of the atomic edifice, the fundamental and malleable piece for infinite constructions. From the seemingly inert mineral to the complex biological models that structure and support life.

At the time of the atomic birth, the hypothesis of a new concept was raised, supported, however, by the basic movement models.

The Neutron, being a structure formed by other particles, could be compared to a mini planetary system, just as the atom would come to be.

At this point, the idea of how all larger structures will develop begins.

As you read this book, it became easier to accept this possibility. It was necessary to go all this way to be able to look back and see this hypothesis more clearly.

By centrifugal force, the nuclear system, the Neutron, threw the electron out of its system and, with that, Hydrogen was born – the first and simplest chemical element.

With the metamorphosis of the Neutron through this birth, a new moment occurs: we will know it as a Proton.

For reasons that we still do not know, this simple atom will receive (or develop) a new Neutron, generating Deuterium. And so, it will continue through the same process until Tritium.

With this progress, another important hypothesis is raised: the transmutation of Hydrogen into Helium.

This possibility generates the idea of Atomic Growth in a natural way without the need for Nuclear Fusion.

These are new and challenging concepts for known models. The mind itself will have great difficulty accepting this information, since the predominant ideas have already crystallized.

I believe that the mental condition of acceptance will be more difficult than the understanding itself.

The path is new and bold and will require great detachment from the reader or researcher.

Returning to the dryness of reason, following this new concept, the atom will grow naturally until it reaches the Uranium phase. The emission of an electron will open the door to the nucleus until it reaches 92 electrons.

Like all transformism, the atom was born, grew and will die due to the instability of the system.

The energy that was concentrated and immobilized in matter tends to return to its dynamic primary form.

The phase of the universe that was compacted is now freeing itself for new flows and possibly new returns to dynamic concentration as matter.

The energy closed in a vortex of concentration by centripetal force, stabilized with the circular movement and opened as a vortex by centrifugal force, returning to its initial condition.

We can see a cycle in this process and this same cyclical model will develop in all the phases to come.

In this coming and going in the form of a vortex of energies, other important and interesting phenomena are generated, Attraction and Repulsion, or the negative (-) and positive (+).

These movements in the form of vortices will be vital for the atomic structuring and for the construction of new combinations, thus generating infinite models and manifestations.

The energy was compressed through concentration to the condition of matter. In reality, solidity is more of a sensation caused by the extremely high speed of the system.

If the energy is concentrated and becomes something with these characteristics, naturally, everything that is generated in this way must have the same nomenclature.

In this case, the concept of antimatter would be more of a nomenclature, due to a differentiated condition of the behavior of the structure of some particle that makes up the atomic building.

The atom and its nucleus are like a mechanical clock, which requires several types of gears to function.

I cite the transition proposed in this work of the Neutron that lost an electron and something else, becoming the Proton. The Proton, in reality, would be like a clock that lost a gear.

The Positron, the positively charged electron, would be one of the gears in the system.

With this in mind, the Positron can also be classified as matter.

In the case of the collision of the Positron with the Electron, two vortices of opposite models, the fit causes the waves to unfold, spreading into light waves and other forms.

Nebulae, in turn, these clusters of atoms, must have come from somewhere, as they are found in various parts of the universe and under different conditions.

The hypothesis that these atoms were generated in these locations by the processes described here is perfectly logical.

With the proposal of atomic growth without the need for nuclear fusion, naturally, heavier elements, known as metals, will be formed over time. This fact may explain the presence of dust in nebulae, even those that do not contain stars (which would make it impossible for exploding stars to be the origin of the dust).

With the formation of stars, the process may be similar to the formation of the atomic nucleus.

Drawing a parallel, in the nucleus there is a union of particles structured by a movement of concentration through a vortex and balanced by a circular movement.

In the case of a star, the path follows the same principle, but the composition will be made up of gases and dust. The rotation or circular movement will maintain the system.

With planets, as already proposed by the hypothesis, the rotation of denser point concentrations within a gaseous mass will cause them to be ejected from the star.

In the atom, the difference in charge keeps the electron bound to the nucleus. In the case of a planet, the phenomenon must be similar. Therefore, a solar system is similar to an atomic system.

The galaxy, drawing a comparison, would be a type of molecule, that is, a structure obeying the same vortex principle in the form of expansion and respecting the same laws of attraction of its components.

In this case, it is easier to analyze by observing its components and their manifestations of expansion.

It is important to note that all these processes have ingredients that keep them together, thus preventing their pulverization.

Following this principle of organization for the expansion of the galaxy to occur, there are, on even larger scales, systems of galaxies. The expanding vortex model is always followed, in which the Curve is the vital path because it requires less work, and is therefore faster and more economical.

The Curved Path is the option for everything that moves in the universe, be it light, matter or other types of waves and mass.

The Black Hole is a gigantic vortex of concentration, which could be described by the negative charge sign (-).

This phenomenon is similar to terrestrial tornadoes, which attract whatever is around them by dragging them.

It should be noted that this model is half of a process that opens and closes on itself. This is what happens in the galaxy, giving balance to the system.

This process is the same as when making a Milkshake.

Finally, the enigmatic Gravitation.

This attraction process occurs in various scales, from the formation of particles to systems of systems of galaxies.

The opening of the system and the return on itself promotes the attraction of matter, sets of matter and waves.

This principle is what keeps the systems together, preventing the pulverization of these sets.

This information can clarify some mysteries that science is debating: Dark Energy and Dark Matter.

I believe that scientists' attention to the hypotheses presented here can contribute to a greater understanding of the universe.

There was also an important comparison between the mechanisms of the microcosm and the macrocosm.

What has been presented here questions the linearity of the expansion of the universe and the forces that maintain the cohesion of structures, such as, for example, galaxies, which do not pulverize.

What is observed through the lenses of telescopes may not be reality, but an illusion, since the waves that reach the detectors may come through curved paths, which, in turn, suggests that the speeds are greater than what would be necessary to cross the actual distance in a straight line.

Although they are also based on already recognized facts, the hypotheses are strongly supported by logic, making this first part of the book more intellectual.

In the second part of this work, the information was based on new scientific observations and worked with all the rigor that science is accustomed to.

The noble collaboration of Dr. Lior Shamir was vital to the completion of this book. In this way, with the information provided by the telescopes and Dr. Shamir's wisdom, we can reach higher conclusive goals.

Despite the strong evidence presented throughout this work, the first part, being more theoretical, requires corroborating facts to strengthen the idea.

It would be very difficult to prove the bending of light, for example, if it were not for the observations of redshift, which demonstrate its possibility at large distances. In short, the light from galaxies that rotate in the opposite direction to the Milky Way is brighter than those that rotate in the same direction.

This bias can be observed and quantified as a fact.

As the arguments in this book suggest, the variation in redshift may be a consequence of the path taken by the light being longer than the apparently linear one.

The bias observed by Dr. Shamir was a key piece in this investigation, as it managed to show what the optical illusion hides.

With this information, we can make important new hypotheses about the cosmological mechanism:

• Light travels a longer path than imagined and detected by instruments.

• Sidereal masses, such as galaxies, stars, quasars and nebulae, may be closer to the observer on Earth than imagined.

• Light may suffer small wear and tear by traveling longer paths, which may increase the red shift.

• This new information undermines the theory that the universe expands linearly beyond the speed of light, when, in fact, its expansion probably occurs in the form of vortices.

I thank you, reader and researcher, for dedicating your precious time to these writings.

Sergio Antonio Meneghetti – Lior Shamir

About the Author

Sergio Antonio Meneghetti

Born in Pindorama – São Paulo – Brazil, married, father of three children.
Intuitive Scientist, Writer, Speaker and Chemist.
Universal Ambassador of Peace – France – Geneva – Switzerland – Cercle Universel des Ambassadeurs de la Paix

Author of the books:

- O Sertanejo de Goiás - Romance Fiction
- Gestão é Uma Arte – Human Management
- The Reconstruction of the Universe - Scientific treatise on the universe and life
- O Fim Sem Fim do Universo - The future of life and the universe
- Intuition Working Tool - Self-development – English version – USA
- Intuição, Ferramenta de Trabalho - Self-development
- O Cavalinho Dourado – Children's Litarature
- Paz no Mundo - Volume I - Poetry
- Paz no Mundo - Volume II - Poetry
- O Pequeno Florista - Children's Litarature
- Liberdade da Consciência - Philosophy
- Vida de Água - Romance Fiction
- A Construção do Pensamento - Philosophy
- Socialmente Falando – Sociology

– Intuição para Mulheres - Self-development
– Sem Saber Sabino - Tales
– Emileão - Children's Litarature
– Multiplicando a Genialidade - Self-development
– Multiplying the Genius Within - English Version
– Homem de Barro
– For Those Who Work in New York - Your biggest strategy
– Do Hidrogênio ao Hélio – Sem Fusão Nuclear

Member of the International Poets Association
Member of the Peace Movement – Poetas Del Mundo
Member of the Franz Liszt Foundation – France

Author of Scientific Hypotheses by Intuitive Psychic Perception:
– Formation of the Subatomic Particle
– Birth of a new planet in our solar system.

Participation in TV, Websites, Radio, national and international magazines.

Jobs:
– Lyondellbasell
– Chevron Química do Brasil
– Institute for Energy and Nuclear Research – IPEN
– EMCA
– Atlas Chemical Industries (Oxiteno)
– SAAB SCANIA

Lior Shamir

Associate Professor- Computer Science - Kansas State University

Ph.D. - 2006, Michigan Technological University
Computational Science & Engineering
M.Sc. - 2003, Open University
Computer Science
B.A. - 1998, Tel Aviv University
Computer Science

Professional experience

Lior Shamir received a Bachelor's degree computer science from Tel Aviv University in 1998. He received an M.S. degree in 2003 from Open University and Ph.D from Michigan Technological University in 2006. He did his postdoc work at the National Institute on Aging of the National Institutes of Health (NIH). He has eight years of experience in the software and technology industry in various positions, from a programmer to a Chief Technology Officer (CTO).

Research

Lior's research focuses on data science. His approach to data science is holistic, combining a broad range of methods in machine learning, soft computing, and computational statistics to develop new paradigms that can turn data into knowledge and scientific discoveries. The other part of his research is the application of the data science methodology to data from a large number of disciplines. So far, he made data-driven discoveries in fields such as astronomy, biology, medicine, humanities (art, music), psychology, marine biology, and more. He takes part in multiple collaborations such as the Large Synoptic Survey Telescope (LSST), the Astrophysics Source Code Library (ASCL), and the Midwest Big Data Hub (MBDH), where he has been serving on the steering committee since it was founded.

In addition to basic and applied research in data science, he also works on advancing inclusion in higher education, making higher education more accessible to underrepresented and underserved populations in STEM.